PROTEIN PRODUCTION
BY BIOTECHNOLOGY

This volume is based upon revised and expanded versions of papers presented at the fourth annual biotechnology symposium of the Biological Council, on the theme 'Protein Production: the exploitation of microorganisms, cells and animals to make useful proteins'.

Other titles in the Elsevier Applied Biotechnology Series:

M. Y. Chisti. *Airlift Bioreactors*

E. J. Vandamme (ed.). *Biotechnology of Vitamins, Pigments and Growth Factors*

PROTEIN PRODUCTION BY BIOTECHNOLOGY

Edited by

T. J. R. HARRIS

Director of Biotechnology, Glaxo Group Research Limited,
Greenford Road, Greenford, Middlesex UB6 OHE, UK

ELSEVIER APPLIED SCIENCE
LONDON AND NEW YORK

ELSEVIER SCIENCE PUBLISHERS LTD
Crown House, Linton Road, Barking, Essex IG11 8JU, England

Sole Distributor in the USA and Canada
ELSEVIER SCIENCE PUBLISHING CO., INC.
655 Avenue of the Americas, New York, NY 10010, USA

WITH 23 TABLES AND 50 ILLUSTRATIONS

British Library Cataloguing in Publication Data

Protein production by biotechnology.
1. Proteins. Synthesis
I. Harris, T. J. R.
547.7'50459

ISBN 1-85166-401-7

Library of Congress Cataloging-in-Publication Data

Protein production by biotechnology/edited by T. J. R. Harris.
    p.   cm.
    Based upon revised and expanded versions of papers presented at the fourth
annual biotechnology symposium of the Biological Council.
    Includes bibliographical references.
    ISBN 1-85166-401-7
    1. Proteins—Biotechnology—Congresses.   I. Harris, T. J. R. (Tim J. R.)
II. Biological Council.
    [DNLM: 1. Biotechnology—congresses.   2. Protein Engineering—congresses.
QU 55 P96043]
TP248.65.P76P76   1990
660'.63--dc20
DNLM/DLC
for Library of Congress   C 366/90   89-17095
                                                 CIP

Phototypesetting by Tech-Set, Gateshead, Tyne & Wear.
Printed in Great Britain by Page Bros (Norwich) Ltd.

PREFACE

There are very few parts of biology that remain free from the influence of Genetic Engineering developed in the early 1970s. Disciplines as wide apart as Brewing, Forensic Science and Population Genetics have all been affected in some way. The major impact, however, has been to create a new science of Biotechnology — a part of which is the production of proteins in a variety of cellular systems. Initially, bacterial systems such as *E. coli* were used but it soon became apparent that this prokaryotic host was not suitable for the preparation of more complicated proteins.

In December 1988, a Symposium sponsored by the Biological Council organised by Dr Chris Hentschel and myself was held at the Middlesex Hospital Medical School in London to discuss alternative methods of protein production and to review some of the applications of the proteins so produced.

The presentations at this meeting form the substance of this book. The theme is apparent from the first part where the expression of proteins and their domains in yeast is described and compared to other fungal and bacterial systems, such as *Aspergillus* and *Bacillus subtilis*. The successful use of recombinant yeast to produce hepatitis B surface antigen for vaccine purposes is particularly pertinent.

There is emphasis in this book on the use of mammalian cells in tissue culture as hosts for making recombinant proteins. These molecules tend to be more like their natural counterparts as far as glycosylation and other postranslational modifications are concerned. Several expression systems are available including infecting cells with recombinant viruses.

The subtlety of the control of gene expression can be probed by analysing the transcription and translation of genes containing upstream sequences that have been transfected into mammalian cells. More sophisticated levels of tissue specific control can be analysed by making transgenic mice containing larger pieces of DNA including far upstream and downstream sequences and DNAse hypersensitivity sites. Transgenic animals can also be used to make proteins in large scale by engineering them to express the proteins in the mammary gland so that they are secreted into milk. All these aspects are covered in the chapters of this book.

Several therapeutic proteins are now available in the market place, including insulin, growth hormone, hepatitis B surface antigen, t-PA (tissue plasminogen activator) and erythropoietin. Some of the problems inherent in using proteins as drugs are addressed — particularly production at large scale and delivery to their site of action. Genetic engineering itself can help, as evident from the extensive modifications done to the t-PA gene, to produce mutant proteins containing fewer domains than the native protein and with potential beneficial effects for fibrinolytic therapy.

This rather direct utilisation of proteins is distinct from their indirect use in drug discovery where the expressed protein is used to aid the finding of a small molecule agonist or antagonist. The cloning and analysis of several transmembrane receptor proteins is a paradigm for activities of this kind and may ultimately become more important. Nevertheless, proteins, such as the colony stimulating factors, are very effective drugs and relatively free from side effects.

It is likely that this book will be of interest to students, graduates and postgraduates in the medical and biological sciences. It does capture, I believe, the essence of the meeting and if, by reading it, students want to undertake Biotechnology as a career instead of some apparently more lucrative alternative, I shall be satisfied. I know that this satisfaction would be shared by Chris Hentschel and Professor Peter Campbell, both of whom helped to make the meeting happen in the first place.

T. J. R. Harris

CONTENTS

LIST OF CONTRIBUTORS

M. ANTONIOU
Laboratory of Gene Structure and Expression, National Institute for Medical Research, The Ridgeway, Mill Hill, London NW7 1AA, UK

P. G. ASHTON-RICKARDT
Department of Molecular Biology, University of Edinburgh, King's Buildings, Mayfield Road, Edinburgh EH9 3JR, UK
Present address: Department of Pathology, University of Edinburgh Medical School, Teviot Place, Edinburgh EH8 9AG, UK

M. BARON
Department of Biochemistry, University of Oxford, South Parks Road, Oxford OX1 3QU, UK

M. M. BENDIG
Medical Research Council, Collaborative Centre, 1–3 Burtonhole Lane, Mill Hill, London NW7 1AD, UK

M. BEVAN
IPSR Cambridge Laboratory, Maris Lane, Trumpington, Cambridge CB2 2JB, UK

J. R. BIRCH
Celltech Ltd, 216 Bath Road, Slough, Berkshire SL1 4EN, UK

M. J. BROWNE
Beecham Pharmaceuticals Research Division, Biosciences Research Centre, Great Burgh, Yew Tree Bottom Road, Epsom, Surrey KT18 5XQ, UK

I. D. CAMPBELL
Department of Biochemistry, University of Oxford, South Parks Road, Oxford OX1 3QU, UK

J. E. CAREY
Beecham Pharmaceuticals Research Division, Biosciences Research Centre, Great Burgh, Yew Tree Bottom Road, Epsom, Surrey KT18 5XQ, UK

C. G. CHAPMAN
Beecham Pharmaceuticals Research Division, Biosciences Research Centre, Great Burgh, Yew Tree Bottom Road, Epsom, Surrey KT18 5XQ, UK

S. H. COLLINS
Delta Biotechnology Limited, Castle Court, Castle Boulevard, Nottingham NG7 1FD, UK

P. COLLIS
Laboratory of Gene Structure and Expression, National Institute for Medical Research, The Ridgeway, Mill Hill, London NW7 1AA, UK

T. M. DEXTER
CRC Department of Experimental Hematology, Paterson Institute & Christie Hospital, Withington, Manchester M20 9BX, UK

N. DILLON
Laboratory of Gene Structure and Expression, National Institute for Medical Research, The Ridgeway, Mill Hill, London NW7 1AA, UK

I. DODD
Beecham Pharmaceuticals Research Division, Biosciences Research Centre, Great Burgh, Yew Tree Bottom Road, Epsom, Surrey KT18 5XQ, UK

J. ERRINGTON
Microbiology Unit, Department of Biochemistry, University of Oxford, South Parks Road, Oxford OX1 3QU, UK

T. FAURE
Transgene, S.A., 11 rue de Molsheim, 67000 Strasbourg, France

D. R. GREAVES
Laboratory of Gene Structure and Expression, National Institute for Medical Research, The Ridgeway, Mill Hill, London NW7 1AA, UK

F. GROSVELD
Laboratory of Gene Structure and Expression, National Institute for Medical Research, The Ridgeway, Mill Hill, London NW7 1AA, UK

O. HANSCOMBE
Laboratory of Gene Structure and Expression, National Institute for Medical Research, The Ridgeway, Mill Hill, London NW7 1AA, UK

J. HURST
Laboratory of Gene Structure and Expression, National Institute for Medical Research, The Ridgeway, Mill Hill, London NW7 1AA, UK

G. ITURRIAGA
IPSR Cambridge Laboratory, Maris Lane, Trumpington, Cambridge CB2 2JB, UK

A. J. KINGSMAN
Department of Biochemistry, University of Oxford, South Parks Road, Oxford OX1 3QU, UK

S. M. KINGSMAN
Department of Biochemistry, University of Oxford, South Parks Road, Oxford OX1 3QU, UK

G. M. P. LAWRENCE
Beecham Pharmaceuticals Research Division, Biosciences Research Centre, Great Burgh, Yew Tree Bottom Road, Epsom, Surrey KT18 5XQ, UK

M. LINDENBAUM
Laboratory of Gene Structure and Expression, National Institute for Medical Research, The Ridgeway, Mill Hill, London NW7 1AA, UK

P. MEULIEN
Transgène S.A., 11 rue de Molsheim, 67000 Strasbourg, France

A. MOUNTAIN
Celltech Ltd, 216 Bath Road, Slough, Berkshire SL1 4EN, UK

K. MURRAY
Department of Molecular Biology, King's Buildings, Mayfield Road, Edinburgh EH9 3JR, UK

A. PAVIRANI
Transgène S.A., 11 rue de Molsheim, 67000 Strasbourg, France

D. B. PRITCHETT
Department of Molecular Neuroendocrinology, Zentrum für Molekulare Biologie (ZMBH), Universität Heidelberg, Im Neuenheimer Feld 382, Postfach 10 62 49, D-6900 Heidelberg, FRG

J. H. ROBINSON
Beecham Pharmaceuticals Research Division, Biosciences Research Centre, Great Burgh, Yew Tree Bottom Road, Epsom, Surrey KT18 5XQ, UK

J. H. SCARFFE
CRC Department of Medical Oncology, Paterson Institute & Christie Hospital, Withington, Manchester M20 9BX, UK

P. H. SEEBURG
Department of Molecular Neuroendocrinology, Zentrum für Molekulare Biologie (ZMBH), Universität Heidelberg, Im Neuenheimer Feld 382, Postfach 10 62 49, D-6900 Heidelberg, FRG

A. E. SMITH
Integrated Genetics Inc., One Mountain Road, Framingham, Massachusetts 01701, USA

G. L. SMITH
Department of Pathology, University of Cambridge, Tennis Court Road, Cambridge CB2 1QP, UK
Present address: Sir William Dunn School of Pathology, Oxford University, South Parks Road, Oxford OX1 7EN, UK

S. STAHL
Biogen S.A., Geneva, Switzerland
Present address: Room 106, Building 2, National Institutes of Health, Bethesda, Maryland 20892, USA

N. STEBBING
ICI Pharmaceuticals, Alderley Park, Macclesfield, Cheshire SK10 4TG, UK

W. P. STEWARD
CRC Department of Medical Oncology, Paterson Institute & Christie Hospital, Withington, Manchester M20 9BX, UK

D. TALBOT
Laboratory of Gene Structure and Expression, National Institute for Medical Research, The Ridgeway, Mill Hill, London NW7 1AA, UK

N. G. TESTA
CRC Department of Experimental Hematology, Paterson Institute & Christie Hospital, Withington, Manchester M20 9BX, UK

E. TOMLINSON
Advanced Drug Delivery Research, Ciba-Geigy Pharmaceuticals, Horsham, West Sussex RH12 4AB, UK

G. TURNER
Department of Microbiology, Medical School, University of Bristol, Bristol BS8 1TD, UK
Present address: Department of Molecular Biology and Biotechnology, University of Sheffield, Sheffield S10 2TN, UK

G. B. VAN ASSENDELFT
Laboratory of Gene Structure and Expression, National Institute for Medical Research, The Ridgeway, Mill Hill, London NW7 1AA, UK

M. VIDAL
Laboratory of Gene Structure and Expression, National Institute for Medical Research, The Ridgeway, Mill Hill, London NW7 1AA, UK

Chapter 1

IS BACILLUS AN ALTERNATIVE EXPRESSION SYSTEM?

JEFFERY ERRINGTON

Microbiology Unit, Department of Biochemistry, University of Oxford, Oxford, UK

&

ANDREW MOUNTAIN

Celltech Ltd, Slough, UK

INTRODUCTION

In this review we consider the possible use of *Bacillus subtilis* as a host for the production of heterologous proteins. There are several potential advantages to be gained from the use of this organism, particularly its efficient secretion of proteins into the growth medium. Although it is unlikely to become the first choice host for the production of certain types of protein, for example potentially therapeutic mammalian proteins, which often undergo specific and necessary post-translation modifications, there are certainly some important groups of proteins, for example industrial enzymes, for which the Bacilli are already heavily used. The market for such enzymes is likely to grow very rapidly as recombinant DNA methods allow for the production of proteins from diverse and often little characterised microorganisms. Exploitation of these natural products will depend upon the development of versatile and robust expression systems, and there is growing evidence that *Bacillus* will be a useful system. On the other hand, *Bacillus* has certain undesirable properties as a host, such as the elaboration of proteases, which can cause loss of product by degradation. If such problems can be overcome, and in this review we assess the technical difficulties involved, then *B. subtilis* may turn out to be the host of choice for the production of many of the new industrial enzymes.

1

Advantages of *Bacillus* as an Expression Host

There are four basic advantages to the use of *Bacillus* as a host for the expression of heterologous proteins.

1. Secretion
Like most of the Bacilli, *B. subtilis* efficiently secretes a variety of proteins directly into the growth medium. From an industrial point of view secretion of the protein product is a major advantage. It allows the accumulation of the native, active product to high yield and in a relatively pure form. Product recovery is thereby greatly simplified compared with high-level intracellular expression, which often leads to the formation of inclusion bodies in which the protein is in an insoluble, inactive form. The saving in expenditure on downstream processing probably accounts for the fact that *Bacillus* exoenzymes, predominantly proteases and amylases, presently comprise about 45% of the commercial enzymes market.[1] However, although the secretion of prokaryotic exoproteins can be very efficient (yields of up to 10 g/litre), there is some evidence that this may not be the case for most eukaryotic exoproteins (see below).

2. Safety
An important factor in host choice is its acceptability with regulatory authorities such as the US Food and Drug Administration (FDA). *Bacillus subtilis* is generally recognised as non-pathogenic and it does not produce endotoxins. Because of this and its long history of safe use in the food and household enzymes industries, it was granted GRAS (Generally Regarded As Safe) status by the FDA. This does not apply to certain other potential microbial hosts such as *E. coli* and is a significant impediment to the latter's use.

3. Genetics
Methods for studying and manipulating genes are arguably more advanced in *B. subtilis* than in any other organism except *E. coli*. Indeed, the natural competence for DNA transformation and the opportunities afforded by efficient uptake and incorporation of linear DNA molecules make this organism even easier to manipulate than *E. coli* in some respects (see below). Thus, vector systems have developed rapidly in *B. subtilis*. As a result of the information accumulated over many years on the genetics and molecular biology of this organism, novel vector and host manipulations are much more predictable and reliable than they are in less well-characterised organisms.

4. Fermentation and Product Recovery Technologies

As already mentioned above, the Bacilli have a long history of industrial use. Consequently, technologies for large-scale fermentation and for product recovery, usually as extracellular soluble protein, are well developed, and they are efficient in commercial terms. The technologies developed should be directly transferable to the production of heterologous proteins by *B. subtilis*.

Disadvantages of *Bacillus*

Problems associated with the use of *B. subtilis* fall into three main categories.

1. Plasmid Instability

Because of the advantages described above and general academic interest in this organism, particularly in its development cycle, attempts to use *B. subtilis* as a host for cloning and recombinant DNA manipulations began as early as the mid-1970s. Unfortunately, for many years there were problems associated with the use of plasmids in *B. subtilis* (reviewed in Ref. 2). Both structural and segregational instabilities were common and the reasons for this are only just becoming clear (see below). This was a major impediment to the development of efficient expression systems in *B. subtilis*: the obvious approach was to model new expression systems on those that proved to be very successful in *E. coli* but these were nearly all plasmid-based!

2. Lack of Suitable Genetic Control Systems

Although *B. subtilis* is one of the best characterised micro-organisms, in comparison with *E. coli* our understanding of the regulation of gene expression must still be regarded as primitive. The highly characterised *lac*, *trp*, and lambda transcriptional control systems of *E. coli* provide powerful, reliable genetic elements that have been utilised in various successful expression systems. Other aspects of the mechanisms controlling gene expression, such as mRNA stability and translational control are also much better understood in *E. coli* than in *Bacillus*. Thus, there are only a few *Bacillus* genes that have been characterised sufficiently to allow their expression signals to be directly utilised in controllable expression systems. It is therefore not yet easy to design expression systems in *Bacillus* and they are likely to be inherently less reliable than those of *E. coli*.

3. Proteolytic Degradation of Products

One of the most important original reasons for interest in the Bacilli was protease production. However, proteases are a major problem in the production of heterologous secreted proteins because of degradation.

DEVELOPMENT OF *BACILLUS SUBTILIS* EXPRESSION SYSTEMS

Given the obvious advantages of using *B. subtilis*, what approaches could be taken to overcome the problems described above?

Stable Maintenance of the Expression System

1. Plasmids

Two classes of recombinant plasmid instability can be distinguished: structural instability, involving mostly DNA deletions but also insertions and other rearrangements; and segregational instability, complete loss of the plasmid. Both of the classes are well documented, and there are various explanations depending on the experimental system being considered. First, there is the use of heterologous plasmids not suited to stable maintenance in *Bacillus*. Until recently, most of the plasmid vectors used in *Bacillus* were derived from *Staphylococcus aureus*. Because of its apparently greater copy number and stability compared with other plasmids, almost all early attempts to develop expression systems used the pUB110 replicon from *S. aureus*. Although the parental plasmid is very stable (no detectable segregation in about 100 generations; A. Mountain, unpublished results), insertions that increase the size of the plasmid dramatically affect its segregational stability, largely through a reduction in copy number.[3] Many of the plasmids of *S. aureus* origin were also shown to generate single-stranded DNA when replicating in *B. subtilis*.[4] Single-stranded DNA has been suggested to initiate recombination,[5] and this may explain some of the structural instability observed. *B. subtilis* also seems to have particularly active recombination systems. In competent cells, certain kinds of recombination event occur up to four orders of magnitude more frequently than in *E. coli*.[6] Also, there is some evidence that insertions of heterologous DNA into some plasmids in *B. subtilis* (particularly those that generate single-stranded DNA during replication) cause impaired termination of replication, leading to the formation of multimeric forms, which are inherently segregationally unstable.[7]

A second explanation for plasmid instability lies in our ignorance of plasmid biology in *B. subtilis* (and in Gram-positive bacteria generally). Many of the constructions involved in developing potential expression systems took little account of the possible effects on plasmid stability through interference with replication or partitioning functions, since these had not been mapped on the plasmid genomes. For example, two regions of pUB110 that have been deleted in hybrids made for a variety of purposes, were recently shown to be required for normal plasmid stability.[8-10]

A third important contribution to plasmid instability arises from the use of uncontrolled expression systems. Strong transcription can affect plasmid stability in two principal ways: firstly, by readthrough into areas involved in replication or partitioning; secondly, by over-expression of a product with a deleterious effect on the host. There is some evidence that the latter situation may even arise by strong expression of secreted proteins during the growth phase in *B. subtilis* (A. Mountain and M. Nugent, unpublished).

The problems of structural and segregational stability can overlap to a certain extent. In situations where the plasmid is (segregationally) destabilised by a reduction in copy number, strong selective pressure would clearly lead to the accumulation of deletions that remove or inactivate the destabilising function.

The obvious way to avoid the problems associated with the use of heterologous plasmids would be to isolate and characterise plasmids from native Bacilli. Several such plasmids are now available (e.g. pTA1060;[11] pTB19;[12] pPOD2000, A. Mountain *et al.*, in preparation). All of the endogenous plasmids described so far seem to have relatively low copy numbers.

An alternative approach would be to manipulate the commonly used plasmids to improve their stability, for example, by introducing functions that stabilise either by improving partitioning at division or by maintaining selection. Although direct selection against plasmid free cells would appear to be an effective solution, in an industrial context a straightforward antibiotic selection, as would be used in the laboratory, is inappropriate on grounds of cost. However, more suitable selective regimes have been devised. The *dal* gene encodes an enzyme, D,L-alanine racemase, that is necessary for the synthesis of D-alanine, which is an essential amino acid involved in cell wall synthesis and is required even during growth in rich media. The plasmid can therefore be maintained by strong selective pressure in a Dal⁻ host.[13] Active partitioning functions of various types have also been introduced with promising results. For example, the *par* or *stab* functions from low copy number indigenous *B. subtilis* plasmids have been shown to stabilise several highly unstable plasmids based on the pUB110 replicon.[14, 15]

2. Chromosomal Integration

An interesting alternative to the use of plasmid replicons is to integrate the expression system into the host chromosome by homologous recombination, taking advantage of the powerful *Bacillus* genetic methods. Two basic integrative strategies are available, distinguished by the apparent number of genetic crossover events that occur between incoming DNA and the host chromosome. Double-crossover events[16, 17] generate extremely stable single copy insertions of the expression cartridge (Fig. 1). Single-crossover 'Campbell' insertions[18] are rather less stable but high copy number, tandemly repeated structures can be generated if a suitable selective marker is incorporated into the repeating unit[19-21] (Fig. 2). Unfortunately, the copy number is sometimes difficult to predict and stability may be a problem.[22] Also, increasing the copy number does not always have a beneficial effect on gene expression (e.g. Ref. 23).

The stable double-crossover insertions may be amplified by introducing extra copies of the expression cartridge at separate chromosomal

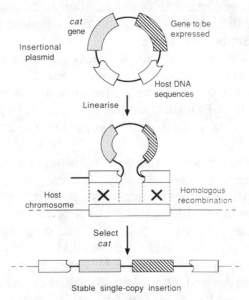

Fig. 1. Integration of an expression cartridge into the *B. subtilis* chromosome by a double-crossover homologous recombination event. The gene to be expressed, together with a selectable marker chloramphenicol acetyl transferase (*cat*) is subcloned into a plasmid incapable of autonomous replication in *B. subtilis* between two fragments of the *B. subtilis* chromosome. After linearising the plasmid, transforming a *B. subtilis* host and selecting for chloramphenicol resistance, the expression cartridge is integrated by a double-crossover event involving the flanking fragments of *B. subtilis* chromosomal DNA.

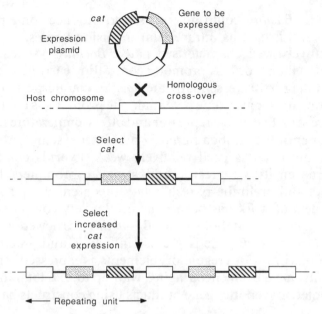

Fig. 2. Integration of an expression cartridge into the *B. subtilis* chromosome by a single-crossover homologous recombination event, and subsequent copy number amplification. The gene to be expressed, together with a selectable marker *cat*, are subcloned into a plasmid incapable of autonomous replication in *B. subtilis* but carrying a single fragment of DNA from the host chromosome. Chloramphenicol resistant transformants arise by a single-crossover recombination event that results in integration of the expression cartridge between directly repeated chromosomal sequences. Further homologous recombination events between the repeated sequences can cause either precise excision of the plasmid by a reversal of the original integration event, or further duplication and amplification of the expression cartridge. Selection for higher levels of antibiotic resistance favours the growth of clones containing multiple tandem repeats.

locations. Although this approach appears promising, it is relatively laborious and it has only been tested for one gene so far.[24]

Controlled Expression

1. Inducible Promoters
Essentially two strategies have been used. The first was to adapt well-characterised controllable promoter systems from *E. coli*. The second was to use native regulatory elements from *Bacillus*. The powerful phage SP01 promoter was combined with the lac repressor/operator to produce the strong IPTG-inducible 'Spac' promoter, which has been used to produce human growth hormone to 25% of the total intracellular protein.[25] Although this was essentially a model system of little commercial importance it illustrates the idea that there is no reason in

principle why *Bacillus* could not be used to produce some proteins in high yield, in this case as intracellular inclusion bodies.

Two relatively well characterised native *Bacillus* systems have also demonstrated their worth. A promoter controlling early gene expression by bacteriophage φ105 (essentially the *Bacillus* counterpart of coliphage lambda) appears to have some potential. The promoter is controlled by a temperature sensitive phage repressor and offers temperature inducibility in the manner of the lambda pL/*cI*857 promoter system.[26] Although the φ105 early promoter is itself relatively weak, hybrids combining the temperature sensitive repressor and stronger promoters have been constructed, and preliminary results have been encouraging.[27]

A great deal of work has been done on the very complex regulatory system for sucrose catabolism in *B. subtilis* (reviewed in Ref. 28). Expression of the *sacB* gene is sucrose inducible and, in combination with mutations in various regulatory elements, is expressed very strongly. Again, results have been encouraging, with up to 25% total intracellular protein quoted for reporter gene products using constructs based on this system.[29]

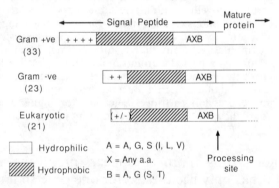

Fig. 3. Comparison of the general structures of signal peptides from Gram-positive, Gram-negative and eukaryotic organisms. The signal peptide consists of three discrete regions: a charged amino terminal, followed by a hydrophobic 'core' and a recognition site for leader peptidases. By comparing data compiled for the leader peptides of many organisms[35, 36] general structures can be derived as shown. Bacterial leader peptides appear to have a much stricter requirement for basic residues at the amino-terminal, with an average of four basic residues in Gram-positive leaders and two in Gram-negative leaders. The hydrophobic core is generally several residues longer for Gram-positive proteins than for Gram-negative or eukaryotic proteins. The leader peptidase recognition site appears to be highly conserved for eukaryotic and Gram-negative organisms at least. Although this conservation is much less clear in the case of Gram-positives, a consensus site is usually present, though sited several residues away from the apparent cleavage site. The distinction may, therefore, be an artefact caused by secondary proteolytic events occurring after translocation and processing by leader peptidase.

As far as secreted proteins are concerned, there are not yet any reports of high yield being achieved using inducible promoters of the types described above. This could be due to saturating the secretion machinery by using promoters that are too strong, or it could reflect the possibility that the secretion machinery operates best in stationary phase. To circumvent the latter problem, growth phase dependent promoters, particularly those from genes encoding secretory proteins, could prove to be useful.

2. Growth Phase Dependent Promoters

The most commonly used growth phase dependent promoters are those of amylase and protease genes. The proteases and amylases are some of the most abundant secreted proteins produced by the Bacilli.[30] Many are commercially important, and hence some of the genes have been characterised in detail. In general terms, these promoters are relatively weak but they are obviously compatible with the secretion machinery. The use of such weak promoters suffers from the disadvantage that relatively protracted fermentations are required, and this inevitably increases the length of exposure of the potential product to proteases, synthesis of which is also largely associated with the stationary phase.

Secretion of Heterologous Proteins

To what extent can secreted proteins from other organisms be utilised by the *Bacillus* secretion machinery? It appears that exoproteins from many other Gram-positive bacteria may be secreted efficiently from *B. subtilis* using their own signal peptides.[31-34] This is not surprising, as signal peptides from Gram-positive organisms tend to have closely related structures. The Gram-positive signal peptide is generally regarded as being longer (average of 33 residues) than its Gram-negative (23 residues) or eukaryotic (21 residues) counterparts (Fig. 3) (data from Refs 35 and 36). However, it is likely that proteolysis of the mature protein after secretion may account for the apparently greater length of at least some Gram-positive signal peptides.[37] Gram-positive and negative signals usually differ in the number of N-terminal basic residues (averages 4 and 2 respectively). The presence of these basic residues is a characteristic feature of prokaryotic signal peptides but they are often absent from eukaryotic signal peptides. These differences probably explain why at least some exoproteins of Gram-negative origin are not secreted efficiently from *Bacillus* (H. Smith, personal communication). Such a problem can sometimes be circumvented by substituting a Gram-

positive signal peptide, such as those from: α-amylase (e.g. Refs 38 and
39), protease (e.g. Refs 40 and 41) and β-lactamase.[42] Unfortunately,
analogous experiments with eukaryotic exoproteins have been relatively
disappointing. Most eukaryotic proteins are not secreted efficiently from
their own signal peptides or from substituted Gram-positive signals (e.g.
Refs 43 and 44). One possible exception is human growth hormone that
has been expressed and secreted to a concentration of about 200 mg/litre
from a vector based on a protease gene.[45] The explanation for the poor
secretion of eukaryotic proteins probably lies in the nature of the
secretion mechanism. It seems that there may be more stringent
structural requirements for membrane translocation of exoproteins in
prokaryotes than in eukaryotes. Possibly, maintenance of a relatively
unfolded conformation before or during secretion is more important in
prokaryotes than in eukaryotes.

A novel approach that shows some promise is based on the use of gene
fusions to the carboxy terminal of efficiently secreted proteins, like α-
amylase. This method has been used to produce a hybrid α-amylase-
human atriopeptin III protein, which was secreted to a concentration of
greater than 1 g/litre. The peptide hormone could be recovered from the
hybrid protein in good yield by a specific proteolytic cleavage.[46]

Protease Production

Undoubtedly the major problem to be overcome if *Bacillus* is to become
an important host for the production of heterologous proteins, is
proteolysis. Several distinct proteolytic activities may be distinguished,
both intracellular and extracellular. Although they are present in only
trace amounts during exponential growth, the activities increase
dramatically in the late growth phase. Two major secreted proteases
make up about 90% of the extracellular proteolytic activity; a neutral
metalloprotease ('NPR') and an alkaline serine protease ('APR', or
'subtilisin'). Genes encoding both of these proteases have been cloned
and then inactivated by several groups.[47-49] However, the residual minor
proteolytic activities remain a problem, particularly with prolonged
fermentations. Some early blocked sporulation mutants exhibit reduced
protease production,[50] and synthesis of at least one of the minor species
of protease has also been abolished using the genetic approach.[51] Thus,
in principle, there is no reason why the other minor proteolytic activities
could not also be removed. However, it is possible that the complete
abolition of proteolysis would have a deleterious effect on the ability of
the cell to maintain growth for the long periods during which the secreted
product would normally accumulate, and hence yields might nevertheless
be limited.

CONCLUSIONS

The Bacilli are already used extensively in the fermentation industry, where they have contributed greatly to the industrial enzymes market. It is likely that the advent of recombinant DNA technology will allow improvements to be made both in the usefulness of existing products, for example, by the introduction of improved thermal resistance properties, and also by the introduction of many new enzymes. The efficient secretion from *B. subtilis* of proteins originating in other Gram-positive organisms using their own signal peptides should ensure that they, at least, will contribute to the extensive use of *Bacillus* as an expression host, provided that the technical problems alluded to above can be overcome. The difficulties in achieving stable maintenance will soon be overcome by using native *Bacillus* replicons, non-antibiotic selections and/or integration into the chromosome. Similarly, the development of suitable systems for controlling gene expression seems to be technically possible, and it seems that soon the only problem remaining will be one of deciding which of the various systems being developed will be most appropriate for a particular product. These developments should permit the widespread use of *B. subtilis* as a host for recombinant enzymes but rapid degradation will probably be more difficult to overcome. In the short term this means that the use of *B. subtilis* for the production of mammalian proteins is likely to be very limited. Further progress will probably depend on greatly extending our knowledge of the secretion mechanism and on detailed characterisation of *B. subtilis* proteases, in particular, the number of different activities, their specificity and their importance for normal cell growth and integrity. The outcome of this work may determine whether *Bacillus* enters general use as the host of choice for the secretion of heterologous proteins, or whether its use remains limited to a small number of proteins that are relatively protease resistant.

ACKNOWLEDGEMENTS

Work in the laboratory of J.E. is supported by the SERC, Celltech Ltd and the Royal Society.

REFERENCES

1. Volesky, B. & Luong, J. H. T., Microbial enzymes: production, purification and isolation. *CRC Crit. Rev. Biotechnol.,* **2** (1985) 119–46.
2. Errington, J., Generalized cloning vectors for *Bacillus subtilis*. In *Vectors: a*

Survey of Molecular Cloning Vectors and their Uses, ed. R. L. Rodriguez & D. T. Denhardt. Butterworths, Boston, 1988, pp. 345–62.

3. Bron, S. & Luxen, E., Segregational instability of pUB110-derived recombinant plasmids in *Bacillus subtilis. Plasmid,* **14** (1985) 235–44.

4. te Riele, H., Michel, B. & Ehrlich, S. D., Single-stranded plasmid DNA in *Bacillus subtilis* and *Staphylococcus aureus. Proc. Nat. Acad. Sci., USA,* **83** (1986) 2541–5.

5. Meselson, M. & Radding, C. M., A general model for genetic recombination. *Proc. Nat. Acad. Sci., USA,* **72** (1975) 358–61.

6. Michel, B. & Ehrlich, S. D., Recombination is a quadratic function of the length of homology during plasmid transformation of *Bacillus subtilis. EMBO J.,* **3** (1984) 2879–84.

7. Gruss, A. & Ehrlich, S. D., Insertion of foreign DNA into plasmids from Gram-positive bacteria induces formation of high-molecular-weight plasmid multimers. *J. Bacteriol.,* **170** (1988) 1183–90.

8. Bron, S., Luxen, E. & Swart, P., Instability of recombinant pUB110 plasmids in *Bacillus subtilis*: plasmid-encoded stability function and effects of DNA inserts. *Plasmid,* **19** (1988) 231–41.

9. McKenzie, T., Hoshino, T., Tanaka, T. & Sueoka, N., The nucleotide sequence of pUB110: some salient features in relation to replication and its regulation. *Plasmid,* **15** (1986) 93–103.

10. Viret, J. F. & Alonso, J. C., Generation of linear multigenome-length plasmid molecules in *Bacillus subtilis. Nucl. Acids Res.,* **15** (1987) 6349–67.

11. Uozumi, T., Ozaki, A., Beppu, T. & Arima, K., New cryptic plasmid of *Bacillus subtilis* and restriction analysis of other plasmids found by general screening. *J. Bacteriol.,* **142** (1980) 315–18.

12. Imanaka, T., Ano, T., Fujii, M. & Aiba, S., Two replication determinants of an antibiotic-resistance plasmid, pTB19, from a thermophilic *Bacillus. J. Gen. Microbiol.,* **130** (1984) 1399–408.

13. Diderichsen, B., A genetic system for stabilization of cloned genes in *Bacillus subtilis*. Bacillus *Molecular Genetics and Biotechnology applications,* ed. A. T. Ganesan & J. A. Hoch. Academic Press, Orlando, 1986, pp. 35–46.

14. Bron, S., Bosma, P., Belkum, M. Van & Luxen, E., Stability function in the *Bacillus subtilis* plasmid pTA1060. *Plasmid,* **18** (1987) 8–15.

15. Chang, S., Chang, S.-Y. & Gray, O., Structural and genetic analysis of a *par* locus that regulates plasmid partitioning in *Bacillus subtilis. J. Bacteriol.,* **169** (1987) 3952–62.

16. Errington, J., A general method for fusion of the *Escherichia coli lacZ* gene to chromosomal genes in *Bacillus subtilis. J. Gen. Microbiol.,* **132** (1988) 2953–66.

17. Niaudet, B., Janniere, L. & Ehrlich, S. D., Integration of linear, heterologous DNA molecules into the *Bacillus subtilis* chromosome: mechanism and use in induction of predictable rearrangements. *J. Bacteriol.,* **163** (1985) 111–20.

18. Duncan, C. H., Wilson, G. A. & Young, F. E., Mechanism of integrating foreign DNA during transformation of Bacillus subtilis. *Proc. Nat. Acad. Sci., USA,* **75** (1975) 3664–8.

19. Albertini, A. M. & Galizzi, A., Amplification of a chromosomal region in *Bacillus subtilis. J. Bacteriol.,* **162** (1985) 1203–11.

20. Janniere, L., Niaudet, B., Pierre, E. & Ehrlich, S. D., Stable gene amplification in the chromosome of *Bacillus subtilis. Gene,* **40** (1985) 47–55.

21. Young, M., Gene amplification in *Bacillus subtilis. J. Gen. Microbiol.,* **130** (1984) 1613–21.
22. Young, M. & Hranueli, D., Chromosomal gene amplification in Grampositive bacteria. In *Recombinant DNA and Bacterial Fermentation,* ed. J. A. Thompson. CRC, Boca Raton, Florida 1988, pp. 157–200.
23. Joyet, P., Levin, D., de Louvencourt, L., Le Reverent, B., Heslot, H. & Aymerich, S., Expression of thermostable alpha-amylase gene under the control of levansucrase inducible promoter from *Bacillus subtilis.* In *Bacillus Molecular Genetics and Biotechnology Applications,* ed. A. T. Ganesan & J. A. Hoch. Academic Press, Orlando, 1986, pp. 479–91.
24. Kallio, P., Palva, A. & Palva, I., Enhancement of α-amylase production by integrating and amplifying the α-amylase gene of *Bacillus amyloliquefaciens* in the genome of *Bacillus subtilis. Appl. Microbiol. Biotechnol.,* **27** (1987) 64–71.
25. Ruppen, M., Band, L. & Henner, D. J., Efficient expression of human growth hormone in *Bacillus subtilis.* In *Bacillus Molecular Genetics and Biotechnology Applications,* ed. A. T. Ganesan & J. A. Hoch. Academic Press, Orlando, 1986, pp. 423–32.
26. Dhaese, P., Hussey, C. & Van Montagu, M., Thermo-inducible gene expression in *Bacillus subtilis* using transcriptional regulatory elements from temperate phage ϕ105. *Gene,* **32** (1984) 181–94.
27. Van Kaer, L., Van Montagu, M. & Dhaese, P., Transcriptional control in the *Eco*RI-F immunity region of *Bacillus subtilis* phage ϕ105. *J. Molec. Biol.,* **197** (1987) 55–67.
28. Klier, A. F. & Rapoport, G., Genetics and regulation of carbohydrate catabolism in *Bacillus. Ann. Rev. Microbiol.,* **42** (1988) 65–95.
29. Zukowski, M. M. & Miller, L., Hyperproduction of an intracellular heterologous protein in a $sacU^h$ mutant of *Bacillus subtilis. Gene,* **46** (1986) 247–55.
30. Priest, F. G., *Extracellular Enzymes,* Van Nostrand Reinhold (UK) Co. Ltd, Wokingham, Berkshire, UK, 1984.
31. Fahnestock, S. R. & Fisher, K. E., Expression of the Staphylococcal protein A gene in *Bacillus subtilis* by gene fusions using the promoter from a *Bacillus amyloliquefaciens* α-amylase gene. *J. Bacteriol.,* **165** (1986) 796–804.
32. Kovacevic, S., Veal, L. E., Hsiung, H. M. & Miller, J. R., Secretion of staphylococcal nuclease by *Bacillus subtilis. J. Bacteriol.,* **162** (1985) 521–8.
33. Soutschek-Bauer, E. & Staudenbauer, W. L., Synthesis and secretion of a heat-stable carboxymethylcellulase from *Clostridium thermocellum* in *Bacillus subtilis. Mol. Gen. Genet.,* **208** (1987) 537–41.
34. Wang, P.-Z., Projan, S. J., Leason, K. R. & Novick, R. P., Translational fusion with a secretory enzyme as an indicator. *J. Bacteriol.,* **169** (1987) 3082–7.
35. Von Heinje, G., Patterns of amino acids near signal-sequence cleavage sites. *Eur. J. Biochem.,* **133** (1983) 17–21.
36. Watson, M. E. E., Compilation of published signal sequences. *Nucl. Acids Res.,* **13** (1984) 5145–64.
37. Takase, K., Mizuno, H. & Yamane, K., NH_2-terminal processing of *Bacillus subtilis* α-amylase. *J. Biol. Chem.,* **263** (1988) 11548–53.
38. Palva, I., Molecular cloning of α-amylase gene from *Bacillus amyloliquefaciens* and its expression in *B. subtilis. Gene,* **19** (1982) 81–7.

39. Shiroza, T., Nakazawa, K., Tashiro, T., Yamane, K., Yanagi, K., Yamasaki, M., Tamura, G., Saito, H., Kawade, Y. & Taniguchi, T., Synthesis and secretion of biologically active mouse interferon-β using a *Bacillus subtilis* α-amylase secretion vector. *Gene,* **34** (1985) 1–8.

40. Vasantha, N. & Thompson, L. D., Secretion of a heterologous protein from *Bacillus subtilis* with the aid of protease signal sequences. *J. Bacteriol.,* **165** (1986) 837–42.

41. Wong, S.-L., Kawamura, F. & Doi, R. H., Use of the *Bacillus subtilis* subtilisin signal peptide for efficient secretion of TEM β-lactamase during growth. *J. Bacteriol.,* **168** (1986) 1005–9.

42. Chang, S., Gray, O., Ho, D., Kroyer, J., Chang, S. Y., McLaughlin, J. & Mark, D., Expression of eukaryotic genes in *Bacillus subtilis* using signals of penP, p. 159–169. In *Molecular Cloning and Gene Regulation in Bacilli,* ed. A. T. Ganesan, S. Chang & J. A. Hoch. Academic Press, New York, 1982.

43. Saunders, C. W., Schmidt, B. J., Mallonee, R. L. & Guyer, M. S., Secretion of human serum albumin from *Bacillus subtilis. J. Bacteriol.,* **169** (1987) 2917–25.

44. Schlein, C. H., Kashiwagi, F., Fujisawa, A. & Weissman, C., Secretion of mature IFN-α-2 and accumulation of uncleaved precursor by *Bacillus subtilis* transformed with a hybrid α-amylase signal sequence-IFN-α-2 gene. *Biotechnology,* **4** (1986) 719–25.

45. Honjo, M., Akaoka, A., Nakayama, A., Shimada, H. & Furutani, Y., Construction of the secretion vector containing the prepro structure coding region of the *Bacillus amyloliquefaciens* neutral protease gene and secretion of *Bacillus subtilis* α-amylase and human interferon β in *Bacillus subtilis. J. Biotechnol.,* **3** (1985) 73–84.

46. Stephens, M., Rudolph, C., Hannett, N., Stassi, D. & Pero, J., Secretion vector for *Bacillus subtilis.* International Patent Application PCT/US8600636, 1986.

47. Fahnestock, S. R. & Fisher, K. E., Protease-deficient *Bacillus subtilis* host strains for production of Staphylococcal protein A. *Appl. Environ. Microbiol.,* **53** (1987) 379–84.

48. Kawamura, F. & Doi, R. H., Construction of a *Bacillus subtilis* double mutant deficient in extracellular alkaline and neutral proteases. *J. Bacteriol.,* **160** (1984) 442–4.

49. Stahl, M. L. & Ferrari, E., Replacement of the *Bacillus subtilis* subtilisin structural gene with an *in vitro*-derived deletion mutation. *J. Bacteriol.,* **158** (1984) 411–18.

50. Piggot, P. J. & Coote, J. G., Genetic aspects of bacterial endospore formation. *Bact. Revs.,* **40** (1976) 908–62.

51. Bruckner, R. & Doi, R. H., Meeting abstract at 4th International Conference on the Genetics and Biotechnology of the Bacilli, San Diego, CA, June 1987.

Chapter 2

EXPRESSION SYSTEMS AND PROTEIN PRODUCTION IN FILAMENTOUS FUNGI

GEOFFREY TURNER*

Department of Microbiology, Medical School, University of Bristol, UK

INTRODUCTION

Filamentous fungi have been of both academic and commercial interest for many years. The last few years have seen the techniques of genetic manipulation extended to a wide range of filamentous fungi. Initially, work was carried out with the genetically well characterised fungi such as *Aspergillus nidulans* and *Neurospora crassa*, but there has been a growing interest in application of these techniques to commercially important fungi. Antibiotics (*Penicillium chrysogenum, Cephalosporium acremonium*), organic acids (*Aspergillus niger, Aspergillus terreus*) and enzymes (*A. niger, A. oryzae, A. awamori, Trichoderma reesei*) are the major economically important products from fungi, and application of molecular techniques is seen as a way of either improving current processes or using the fungi as hosts for production of heterologous proteins. Fungi already used for pharmaceutical or food products are seen to have a commercial advantage for heterologous protein production in that they enjoy the GRAS (Generally Regarded As Safe) status from the US Food and Drug Administration (FDA).

Proteins produced commercially from fungi are generally those which are excreted, and enzymes such as glucoamylase and cellulases are claimed to reach relatively high concentrations in the growth medium (5–40 g crude protein/litre for production strains; see, for example Ref. 1). For a variety of reasons (e.g. extraction and solubilisation problems with internally overproduced proteins), secretion of heterologous proteins

*Present address: Department of Molecular Biology and Biotechnology, University of Sheffield, UK

15

from a microbial host is seen as an important aim, and any micro-organism with a reputation for a high level of protein secretion therefore becomes worthy of further investigation.

INSERTING GENES INTO FUNGI

A number of reviews deal with fungal transformation,[2-4] but an outline is given here. Transformation of fungi by exogenous DNA requires prior removal of the cell wall using lytic enzymes in the presence of an osmotic stabiliser, and the formation of protoplasts. A number of commercial preparations are available for this. DNA uptake is then stimulated by polyethylene glycol in the presence of Ca^{2+} ions. Transformants are most conveniently selected by including a prototrophic nutritional gene in the transforming DNA, and a mutation in the recipient strain. The major drawback with this approach is that the correct mutation has to be introduced into the recipient fungus. When working with genetically well-studied fungi, such as *Aspergillus nidulans* and *Neurospora crassa*, this presents few problems. Since few well-characterised mutants are available in industrial species, this approach may not be so convenient. As a result, a number of antibiotic resistance markers have been developed. These usually consist of a resistance gene of bacterial origin coupled to a fungal promoter, though there are examples of fungal-derived resistance genes. Such markers can be used to transform prototrophic strains directly; some examples are given in Table 1.

In most cases, the transforming DNA integrates into the fungal genome, by both homologous and non-homologous recombination, in single or multiple copies. The latter phenomenon is of considerable importance for increasing gene expression. Multicopy, autonomously replicating plasmids are virtually unknown in filamentous fungi, though some instances of autonomous replication have been documented.[5] Integration can be by homologous and non-homologous recombination with the fungal genome, and the relative proportion of these type of events is species dependent. The multiple integration can be tandem or at many sites, but another biotechnologically important feature is that most transformants are stable during vegetative growth, such that selective pressure is not required to maintain the multicopy state.

FUNGAL PROMOTERS

The transformation systems outlined above enable any gene to be inserted into a fungus in a stable, multicopy state. For some applications

Table 1
Some Selectable Markers Used in Fungal Transformation

Marker	Gene product	Origin	Recipient	Reference
1. Requiring an auxotrophic recipient				
pyr4	Orotidylate decarboxylase	*N. crassa*	*A. nidulans*	26
pyrG		*A. niger*	*A. oryzae*	27
argB	Ornithine transcarbamylase	*A. nidulans*	*A. nidulans*	28
argB		*A. nidulans*	*A. niger*	29
2. Direct transformation of prototrophic strains				
amdS	Acetamidase	*A. nidulans*	*A. nidulans*	30
			A. niger	31
			P. chrysogenum	32
oliC31	ATP synthase subunit 9 (oligomycin resistant)	*A. nidulans*	*A. nidulans*	33
oliC3		*A. niger*	*A. niger*	9
oliC13		*P. chrysogenum*	*P. chrysogenum*	34
benA	Betatubulin (benlate resistant)	*N. crassa*	*N. crassa*	35
hph	Hygromycin phosphotransferase	bacterial	*A. niger*	36
phl	Phleomycin resistance	bacterial	*P. chrysogenum*	11

it may be desirable to express a gene from one fungus in another. The growing literature on fungal genes now provides some general rules for guidance. Within the filamentous Ascomycetes, and their Deuteromycete (asexual) relatives, fungal promoters seem to have a wide host range. *Saccharomyces cerevisiae* is outside this range, possibly because of a greater difference in promoter organisation or problems with intron recognition and splicing. Heterologous non-fungal genes therefore require the addition of a suitable fungal promoter. Alteration of expression of a fungal gene may also be desirable, for instance, where a change in promoter strength or regulation is required.

A number of fungal promoters have been employed in construction of expression systems, though information on the relative strength of these promoters is limited. As in yeast, promoters from enzymes of the Embden–Meyerhoff pathway, such as glyceraldehyde phosphate dehydrogenase and phosphoglycerate kinase, have been isolated and tested. Although the fungi in question are obligate aerobes, and do not use the pathway in a fermentative mode as does yeast, these promoters do appear to be relatively strong in *A. nidulans*.[6,7] Another promoter of similar strength is that of the *oliC* gene of *A. nidulans*[8] and *A. niger*,[9] which encodes mitochondrial ATP synthetase subunit 9.

One approach to comparing promoter strength has been the use of gene fusions with the *E. coli* beta-galactosidase gene, *lacZ*, and a useful set of vectors was constructed for this purpose.[10] These vectors are designed for translational fusions and assay for beta-galactosidase activity is therefore a measurement of protein expression for one particular heterologous system. The *argB* gene of *A. nidulans* is included in the vectors for selection and insertion of single copies of a translational fusion at the host *argB* locus by homologous recombination. While a major class of transformant is the latter, multicopy/multisite insertions are also relatively common, and lead to much higher levels of beta-galactosidase activity. With a moderate to strong promoter, the increased level of betagalactosidase protein is detectable as a stained band in total soluble protein extracted from mycelium[11] (Fig. 1).

The promoters discussed so far result in constitutive expression of

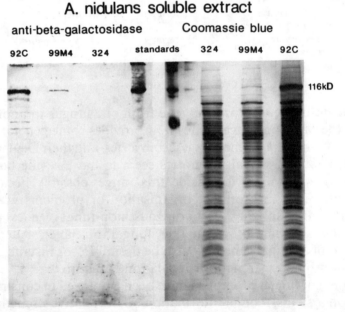

Fig. 1. Expression of an *oliC*/beta-galactosidase translational fusion in *Aspergillus nidulans*. 324, recipient (control) strain; 99M4, single copy fusion gene; 92C, multiple copy fusion gene (approximately 10 copies). The fusion genes were inserted by transformation using the *argB* selectable marker. *E. coli* beta-galactosidase (standard) and the fusion protein are detected as 116kd bands (S. Kerry-Williams, J. Wilding & G. Turner, unpublished).

homologous and heterologous gene products. In order to exert control over expression, other promoters have been employed. One of the three alcohol dehydrogenase enzymes of *A. nidulans, alcA*, is strictly regulated.[12, 13] When grown on glucose or sucrose, the activity is repressed, and full induction occurs when the ethanol replaces carbohydrate as carbon source. Threonine also acts as inducer, and may be used instead of ethanol. The *alcA* gene is positively regulated by a product of the *alcR* gene, and attempts at improving protein yield by inserting multiple copies of gene fusions employing the *alcA* 5' region revealed that the *alcR* product soon becomes rate limiting.[14] Thus additional copies of the *alcA* promoter have to be provided with an increased supply of the *alcR* product to achieve maximum expression. Similar problems have not been reported with constructs using the *Aspergillus niger gla* (glucoamylase) 5' region (see below). Constructs based on the *alcA* promoter of *A. nidulans* did not function in *A. niger* (D. I. Gwynne, personal communication).

Since certain industrial fungi produce large amounts of specific enzymes, it is to be expected that the genes encoding these enzymes would provide strong promoters. The glucoamylase promoter of *A. niger*,[15, 16] the cellobiohydrolase I promoter of *Trichoderma reesei*[1] (A. Harkki, personal communication) and the alpha-amylase (Taka amylase) promoter of *Aspergillus oryzae*[17] have been used in a number of constructs.

IMPROVEMENTS IN HOMOLOGOUS GENE EXPRESSION

Many commercially important fungi have already been subjected to rounds of mutagenesis and selection in order to improve productivity,[18] and it remains to be seen whether molecular approaches can achieve further improvements. Insertion of multiple copies of the amyloglucosidase gene into wild-type strains of *A. niger* have resulted in significant increases in protein yield.[19]

To find out which approach might be effective on a current production strain, molecular analysis of such production strains can be very informative. Smith *et al.*[20] examined a high titre strain of *Penicillium chrysogenum* in this way. One of the key enzymes in the penicillin biosynthetic pathway is isopenicillin N synthetase. Using the gene for this enzyme as a probe, it was observed that gene copy number and steady state mRNA level are greater than in the wild-type strain by 10- and 50-fold respectively, indicating *both* gene amplification and up-promoter mutations. The alpha-amylase promoter used by Christensen *et al.*[17] was isolated from an overproducing strain of *Aspergillus oryzae*.

A further example of where genetic manipulation of homologous gene expression could be valuable is *Trichoderma reesei*. Digestion of cellulose requires a range of enzymes, including cellobiohydrolases and endo-glucanases.[21] Improvements in the efficiency of cellulose digestion might be possible by altering the composition of the cellulolytic mixture.

EXPRESSION AND SECRETION OF FOREIGN PROTEINS

While the means are now available for manipulation of native fungal proteins, a more ambitious aim is to use filamentous fungi as hosts for the production and excretion of foreign proteins.

An obvious approach is to utilise the whole 5' region of a gene encoding a secreted fungal protein such that it includes promoter/regulatory elements, secretion and processing signals, and fuse this to the gene for the desired product. Glucoamylase, alpha-amylase and cellobiohydrolase genes have been used in this way. A synthetic secretion signal has also been used.[16] Since the specificity of secretion signals appears to be rather low, fusion genes utilising fungal promoter and mammalian secretion and processing signals appears to be just as effective.

Cullen *et al.*[15] reported controlled expression and secretion of active bovine chymosin from *A. nidulans* using either the *A. niger* glucoamylase or preprochymosin signal peptide sequences fused directly to prochymosin sequence. Other constructs, which included the glucoamylase propeptide sequence and part of the mature glucoamylase sequence, gave poorer secretion (Table 2). Although the glucoamylase gene was from *A. niger*, expression of the fusion genes was transcriptionally regulated by carbon source in *A. nidulans*, induced in the presence of starch and not with xylose. At low pH of the growth medium, the prochymosin appeared to undergo the normal autocatalytic cleavage to chymosin.

The most successful of these constructs have now been inserted into *A. awamori*, giving similar levels of secreted product.[23]

Gwynne *et al.*[16] reported expression and secretion of human interferon alpha 2 from *A. nidulans* under the control of the *alcA* promoter and regulatory sequences. Secretion was mediated by provision of a synthetic 'consensus' signal sequence.

Human tissue plasminogen activator has also been expressed and secreted from *A. nidulans*.[22] The triose phosphate isomerase 5' region was linked directly to the human tissue plasminogen activator (t-PA) cDNA, which included the natural signal sequence. The active, secreted product resembled authentic mammalian t-PA, having been correctly processed to give the two forms (glycine and serine N-termini) produced by

Table 2

Fusions Used for Secretion of Prochymosin from *Aspergillus nidulans* and *Aspergillus awamori*[a]

GLA PROMOTER---GLA SIGNAL PEPTIDE---PROCHYMOSIN
GLA PROMOTER---GLA SIGNAL PEPTIDE--GLA PROPEPTIDE-------------------------PROCHYMOSIN
GLA PROMOTER---CHYMOSIN SIGNAL PEPTIDE-------------------------------------PROCHYMOSIN
GLA PROMOTER---GLA SIGNAL PEPTIDE--GLA PROPEPTIDE--11 AMINO ACIDS GLA------PROCHYMOSIN

[a]The fusions were introduced using the *pyr4* gene of *Neurospora crassa* in the vector as a selectable marker.[15,23]
GLA: glucoamylase gene from *Aspergillus niger*.

mammalian cells. Glycosylation of the protein had also occurred, but not hyperglycosylation.

PROBLEMS AND SOLUTIONS

Level of Expression

In most cases, heterologous (non-fungal) genes are not expressed as well as the homologous fungal gene, and overcoming this problem is essential if fungi are to be of any commercial value as producers of heterologous proteins. Reduced expression may occur as a result of transcription, translation and/or secretion problems, but the latter seems to be the main barrier. Work with a number of model systems has indicated that efficient production of non-secreted soluble heterologous proteins internally is relatively easy to achieve. However, expression/secretion of mammalian gene products was very inefficient in the 'first generation' strains, being around $100 \mu g/g$ dry weight mycelium. Attempting to account for the low level of expression, Ward[23] observed that the glucoamylase promoter–prochymosin fusion was transcribed efficiently, suggesting problems at subsequent steps in expression.

The secretion pathway in eukaryotes is complex, and very little fundamental study of the genetics or biochemistry of the pathway has been carried out in filamentous fungi. Since such a study would be uneconomic and time consuming for individual companies, short cuts have been attempted. The most detailed published accounts of such attempts are those from Genencor Inc.,[23] who are aiming for commercial production of bovine chymosin by *Aspergillus awamori*, a close relative of *A. niger*.

Firstly, a large number of transformants obtained with the vectors carrying the gene fusions were screened, and a wide variation in yield was seen, which did not clearly correlate with copy number of integrated construct. Secondly, rounds of random mutation and screening led to derived strains with better yields.[24]

The difficulty of obtaining high levels of secreted mammalian proteins may relate to fundamental differences in the secretion pathways of animals and fungi. It should be noted that success has been achieved in the secretion of heterologous fungal proteins. Christensen *et al.*[17] reported secretion of *Rhizomucor meihei* aspartic proteinase (used as a substitute for rennin) from *Aspergillus oryzae* at 3 g/litre when driven by an improved alpha-amylase promoter.

Protease Activity

Since filamentous fungi produce a range of secreted enzymes, including proteases, the desired protein product is susceptible to degradation as it is released into the growth medium. In some fungi, including *A. nidulans*, the production of proteases can be partly repressed by ammonia, which is the preferred nitrogen source for the fungus. This repression is not so effective in *A. niger*, and other steps are necessary. Although random mutagenesis and screening might offer an approach, the fact that fungi often produce more than one protease may be a problem. An alternative approach has involved cloning the gene for a major acid protease, Aspergillopepsin A, from *Aspergillus awamori*.[25] The host protease was subsequently deleted by a transformation and gene replacement strategy. Although some protease activity still remained, growth conditions were modified to reduce this, and further yield improvements for chymosin achieved.[23]

Glycosylation

Few detailed studies have been carried out on the nature of the glycosylation of secreted foreign proteins by filamentous fungi. Upshall *et al.*[22] reported differential glycosylation of human tissue plasminogen activator (t-PA) to give two forms, which resembled those produced by mammalian cells. Christensen *et al.*,[17] however, found that the *Rhizomucor miehei* aspartic proteinase was slightly overglycosylated (N-glycosylation), though this did not affect the activity.

CONCLUDING REMARKS

Whether filamentous fungi can emerge as useful hosts for the expression of mammalian proteins may be decided within the next few years, but improved production of fungal enzymes by fungi seems less of a hurdle. The observation that promoters are often interchangeable within classes of filamentous fungi has been followed by examples of efficient secretion of heterologous *fungal* proteins. Furthermore, secreted fungal proteins are less susceptible to secreted host proteases than are mammalian proteins.

It will be interesting to see how far the mutation and screening approach can be used to obtain further improvements in the secretion of mammalian proteins, and to remove undesirable protease activity.

REFERENCES

1. Knowles, J., Lehtovaara, P., Pentilla, M., Teeri, T., Harkki, A. & Salovuori, I., The cellulase genes of *Trichoderma. Antonie van Leeuwenhoek J. Microbiol.,* **53** (1987) 335–41.
2. Turner, G. & Ballance, D. J., Transformation of *Aspergillus nidulans.* In *Genetic Manipulation of Filamentous Fungi,* ed. J. W. Bennett & L. Lasure. Academic Press, New York, 1985, pp. 259–78.
3. Rambosek, J. A. & Leach, J., Recombinant DNA in filamentous fungi: progress and prospects. *CRC Crit. Rev. Biotechnol.,* **6** (1987) 357–93.
4. Fincham, J. R. S., Transformation in fungi. *Microbiol. Rev.,* **53** (1989) 148–70.
5. van Heeswijck, R., Autonomous replication of plasmids in *Mucor* transformants. *Carlsberg Res. Comm.,* **51** (1986) 433–43.
6. Clements, J. M. & Roberts, C. F., Molecular cloning of the 3-phosphoglycerate kinase (PGK) gene from *Aspergillus nidulans. Curr. Genet.,* **9** (1985) 293–8.
7. Punt, P. J., Dingemans, M. A., Jacobs-Mijesing, B. J. M., Pouwels, P. H. & van den Hondel, C. A. M. J. J., Isolation and characterization of the glyceraldehyde-3-phosphate dehydrogenase gene of *Aspergillus nidulans. Gene,* **69** (1988) 49–57.
8. Ward, M. & Turner, G., The ATP synthase subunit 9 gene of *Aspergillus nidulans*: sequence and transcription. *Molec. Gen. Genet.,* **205** (1986) 331–8.
9. Ward, M., Wilson, L., Carmona, C. & Turner, G., The *oliC3* gene of *Aspergillus niger*: isolation, sequence, and use as a selectable marker. *Curr. Genet.,* **14** (1988) 37–42.
10. van Gorcom, R. F. M., Punt, P. J., Pouwels, P. J. & van den Hondel, C. A. M. J. J., A system for the analysis of expression signals in *Aspergillus. Gene,* **48** (1986) 211–17.
11. Kolar, M., Punt, P. J., van den Hondel, C. A. M. J. J. & Schwab, H., Transformation of *Penicillium chrysogenum* using dominant selection markers and expression of an *Escherichia coli lacZ* fusion gene. *Gene,* **62** (1988) 127–34.
12. Pateman, J. H., Doy, C. H., Olson, J. E., Norris, U., Creaser, E. H. & Hynes, M. J., Regulation of alcohol dehydrogenase (ADH) and aldehyde dehydrogenase (AldDH) in *Aspergillus nidulans. Proc. Roy. Soc. Lond.,* **B217** (1983) 243–64.
13. Gwynne, D. I., Buxton, F. P., Sibley, S., Davies, R. W., Lockington, R. A., Scazzocchio, C. & Sealy-Lewis, H. M., Comparison of the *cis*-acting control regions of two coordinately controlled genes involved in ethanol utilization in *Aspergillus nidulans. Gene,* **51** (1987) 205–16.
14. Gwynne, D. I., Buxton, F. P., Gleeson, M. A. & Davies, R. W., In *Protein Purification: Micro to Macro,* ed. R. Burgess, Alan R. Liss, New York, 1987, pp. 355–65.
15. Cullen, D., Gray, G. L., Wilson, L. J., Hayenga, K. J., Lamsa, M. H., Rey, M. W., Norton, S. & Berka, R. M., Controlled expression and secretion of bovine chymosin in *Aspergillus nidulans. Biotechnology,* **5** (1987) 369–76.
16. Gwynne, D. I., Buxton, F. P., Williams, S. A., Garven, S. & Davies, R. W., Genetically engineered secretion of active human interferon and a bacterial endoglucanase from *Aspergillus nidulans. Biotechnology,* **5** (1987) 713–19.

17. Christensen, T., Woeldike, H., Boel, E., Mortensen, S. B., Hjortshoej, K., Thim, L. & Hansen, M. T., High level expression of recombinant genes in *Aspergillus oryzae. Biotechnology,* **6** (1988) 1419–22.
18. Rowlands, R. T., Industrial strain improvement: mutagenesis and random screening procedures. *Enzyme Microb. Technol.,* **6** (1984) 3–10.
19. Finkelstein, D. B., Rambosek, J. A., Leach, J., Wilson, R. E., Larson, A. E., McAda, P. C., Soliday, C. L. & Ball, C., Genetic transformation and protein secretion in industrial filamentous fungi. In *Fifth International Symposium on the Genetics of Industrial Microorganisms,* ed. M. Alacevic, D. Hranueli & Z. Toman, GIM'86, Jugoslavia. Pliva, Zagreb, 1986, pp. 101–11.
20. Smith, D. J., Bull, J. H., Edwards, J. & Turner, G., Amplification of the isopenicillin N synthetase gene in a strain of *Penicillium chrysogenum* producing high levels of penicillin. *Molec. Gen. Genet.,* **216** (1989) 492–7.
21. Henrissat, B., Driguez, H., Viet, C. & Schulein, M., Synergism of cellulases from *Trichoderma reesei* in the degradation of cellulose. *Biotechnology,* **3** (1988) 722–6.
22. Upshall, A., Kumar, A. A., Bailey, M. C., Parker, M. D., Favreau, M. A., Lewison, K. P., Joseph, M. L., Maroganore, J. M. & McKnight, G. L., Secretion of active human tissue plasminogen activator from the filamentous fungus *Aspergillus nidulans. Biotechnology,* **5** (1987) 1301–4.
23. Ward, M., Chymosin production in *Aspergillus.* In *Molecular Industrial Mycology: Systems and Applications,* ed. S. A. Leong & R. M. Berka. Marcel Dekker, 1989, in press.
24. Lamsa, M. & Bloebaum, P., Mutation and screening to increase chymosin yield in a genetically-engineered strain of *Aspergillus awamori. J. Industr. Microbiol.* (in press).
25. Berka, R. M., Ward, M., Wilson, L. J., Hayenga, K. J., Fong, K. K., Carlomagno, L. P. & Thompson, S. A., Molecular cloning and deletion of the Aspergillopepsin A gene from *Aspergillus awamori. Gene* (submitted).
26. Ballance, D. J., Buxton, F. P. & Turner, G., Transformation of *Aspergillus nidulans* by the orotidine-5'-phosphate decaboxylase gene of *Neurospora crassa. Biochem. Biophys. Res. Commun.,* **112** (1983) 284–9.
27. Mattern, I. E., Pouwels, P. H. & van den Hondel, C. A. M. J. J., Transformation of *Aspergillus oryzae* using the *A. niger pyrG* gene. *Molec. Gen. Genet.,* **210** (1987) 460–1.
28. John, M. A. & Peberdy, J. F., Transformation of *Aspergillus nidulans* using the *argB* gene. *Enzyme Microb. Technol.,* **6** (1984) 386–9.
29. Buxton, F. P., Gwynne, D. I. & Davies, R. W., Transformation of *Aspergillus niger* using the *argB* gene of *Aspergillus nidulans. Gene,* **37** (1985) 207–14.
30. Tilburn, J., Scazzocchio, C., Taylor, G. G., Zabricky-Zissman, J. H., Lockington, R. A. & Davies, R. W., Transformation by integration in *Aspergillus nidulans. Gene,* **26** (1983) 205–21.
31. Kelly, J. M. & Hynes, M. J., Transformation of *Aspergillus niger* by the *amdS* gene of *Aspergillus nidulans. EMBO J.,* **4** (1985) 475–9.
32. Beri, R. K. & Turner, G., Transformation of *Penicillium chrysogenum* using the *Aspergillus nidulans amdS* gene as a dominant selective marker. *Curr. Genet.,* **11** (1987) 639–41.
33. Ward, M., Wilkinson, B. & Turner, G., Transformations of *Aspergillus nidulans* with a cloned, oligomycin-resistant ATP synthase subunit 9 gene. *Mol. gen. Genet.,* **202** (1986) 265–70.

34. Bull, J. H., Smith, D. & Turner, G., Transformation of *Aspergillus nidulans* with a cloned, oligomycin-resistant ATP synthetase subunit 9 gene. *Curr. Genet.,* **13** (1988) 377–82.
35. Orbach, M. J., Porro, E. B. & Yanofsky, C., Cloning and characterisation of the gene for beta-tubulin from a benomyl-resistant mutant of *Neurospora crassa* and its use as a dominant selectable marker. *Molec. Cell Biol.,* **6** (1986) 2452–61.
36. Punt, P. J., Oliver, R. P., Dingemanse, M. A., Pouwels, P. H. & van den Hondel, C. A. M. J. J., Transformation of *Aspergillus* based on the hygromycin B resistance marker from *Escherichia coli. Gene,* **56** (1987) 117–24.

Chapter 3

GENETIC ENGINEERING APPLIED TO THE DEVELOPMENT OF VACCINES

K. MURRAY,[a] S. STAHL[*b] & P. G. ASHTON-RICKARDT[‡a]

[a]*Department of Molecular Biology, University of Edinburgh, Edinburgh, UK*
[b]*Biogen S.A., Geneva, Switzerland*

INTRODUCTION

Man and other animals survive the relentless assault of a large array of microbial pathogens because they have evolved mechanisms that enable them not only to recover from initial attacks and purge themselves of the offender, but also to remember sufficient information about the molecular morphology of an individual aggressor that they are able to repulse further intrusions very quickly and effectively. Through this combination of exposure to a pathogen and clearance and recollection of it, the animal becomes immune to that particular pathogen.

Vaccination is dependent upon the memory of the immune system and seeks to add a particular piece of information to the animal's immunological database without exposing the recipient to the pathogenic effects attending normal infections. Some very effective vaccines against viruses have been made by traditional procedures involving killed preparations of the virus or attenuated or avirulent forms of the virus, which may arise during repeated passage of the wild-type virus through animal hosts or cell lines in culture. Inoculation of an individual with such preparations will normally prime or stimulate the immune system to counter subsequent exposure to the wild-type virus, as this will have the same appearance as the immunizing preparation. In preparing such vaccines killing of the virus culture must obviously be complete and the

*Present address: National Institutes of Health, Bethesda, Maryland, USA
‡Present address: Department of Pathology, University of Edinburgh Medical School, Edinburgh, UK.

avirulent strain unable to revert to virulent. Many serious diseases have been controlled in this way, and smallpox has been essentially eliminated by virtue of the World Health Organization vaccination programme.

A virion lacking its genome would offer a particularly safe means of vaccination if it could be made, but particles or shapes simulating the viral surface offer an equivalent means of stimulating immunological memory without the potential, however remote, for pathogenic effects inherent in the use of a complete virus. Preparations of particulate envelope proteins or surface antigens, are often referred to as 'subunit vaccines'. An outstanding successful subunit vaccine is that developed against hepatitis B virus (HBV) from extensively purified preparations of the surface-antigen particles of the virus from the plasma of infected individuals.[1]

The general area of vaccine development was clearly an early and attractive choice for the application of genetic engineering.[2] The wide range of systems for gene cloning and expression together with the development of rapid methods for nucleotide sequence determination and for oligonucleotide synthesis, and with them opportunities for introduction of mutations at any desired position in a DNA molecule, have provided the means of manipulating genomes with relative ease. In principle these methods could be applied in the development of attenuated live vaccines (which have important practical advantages in ease of application and low cost) once the molecular basis of virulence is understood. This topic is in its infancy, but the recent identification of a single nucleotide change in a non-coding region of the polio type III virus genome that distinguishes virulent and avirulent forms of the virus provides encouragement for such approaches.[3] However, the expression of viral gene products in microbial or animal cells in culture offers an alternative to killed or attenuated viral preparations as a source of raw material for vaccine formulation. This is because recombinant DNA technology has now developed to a stage where cloning and expression of genes are commonplace, and with sufficient application there are probably few proteins that cannot be produced in quite large quantities and in a high degree of purity, although actual levels of expression and problems attending purification of any individual protein vary widely. These procedures have now been used to produce a number of viral and other antigens for vaccine formulations and effective vaccines against hepatitis B made in this way have been approved and are in widespread use.

Surface antigens of viruses have been the principal focus of most attempts to make viral vaccines by genetic manipulation approaches, but antibody formation to other components is well known and the significance of the immunological response to internal viral proteins, particularly in relation to the cell mediated response, is being increasingly recognized. Hepatitis B virus will be used here to illustrate both the successful development of a viral envelope protein into a vaccine against a major disease worldwide (and indirectly against liver cancer), and the importance of other viral antigens in immunity and their potential application in the control of infection.

HEPATITIS B VIRUS

The causative agent of hepatitis B, or serum hepatitis, is a roughly spherical virion with a diameter of 42 nm, often referred to as the Dane particle,[4] which is found in the liver and blood of acutely infected individuals. It is transmitted sexually or parenterally and poses a major threat to public health worldwide for a carrier state may follow infection and the long term sequelae can then include cirrhosis and development of primary liver cancer. The extensive literature on clinical and molecular biological aspects of hepatitis B is summarized in a number of recent reviews and symposia volumes.[5-10] The virus has a very narrow host range, infecting only man and a few other higher primates, of which the chimpanzee is the best known and has been widely adopted for investigational purposes.

The human virus may be regarded as the prototype of a family of viruses now known as the Hepadna Viridae, for similar viruses have been found in a number of species, but in each case the virus is narrowly confined to a specific host. In the blood of infected individuals the virus is usually accompanied by a large excess of particles of its capsid protein, or surface antigen (HBsAg), which may be in the form of 22 nm diameter spherical particles or as filaments. Treatment of the virus with mild detergent removes the envelope to leave a nuclear capsid or core particle with a diameter of 27 nm; such particles are also found in the liver of infected individuals but not in the blood. HBsAg particles consist of two principal components, one (gp27) being a glycosylated form of the other (p24), but larger polypeptides with amino acid sequences overlapping those of these two smaller components (and having the same carboxy terminus) have now been recognized as authentic minor components of both the spherical and filamentous forms of the surface-antigen particles and of Dane particles.[11]

The nucleocapsid or core particles have one major protein component, the core antigen (HBcAg), which surrounds the viral genome, a small

DNA molecule of rather unusual structure. A DNA-dependent DNA polymerase activity is associated with the virus and the core particles, and another viral antigen, the e antigen (HBeAg), is found in various forms, but has been shown to be a derivative of the core antigen.[12] The virus and its components are represented diagrammatically in Fig. 1, which also illustrates the manifestation of infection by the virus in terms of appearance in serum of antigens and antibodies to them in both acute and chronic infections.[8]

PRODUCTION OF HBV ANTIGENS IN MICROBIAL CELLS

The limited host range of the virus means that HBV cannot be propagated in laboratory animals and until very recently it could not be produced in cell culture,[13] which, given the clinical importance of the virus, made it an early candidate for molecular cloning with particular respect to the production of antigens for both diagnostic and vaccine development. Given these objectives, our initial cloning strategy was based upon the insertion of fragments of the viral DNA into the

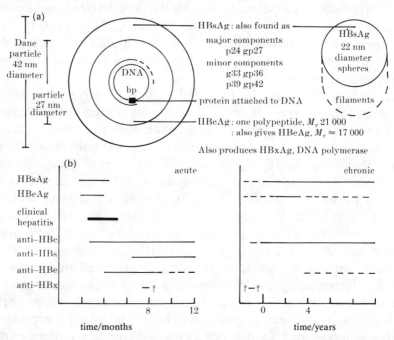

Fig. 1. (a) The components of HBV particles. (b) The appearance of HBV antigens, and antibodies to them, in serum during acute and chronic infections. There are wide variations in the time at which the various markers are found in individual cases; this is indicated in part by the broken lines.

β-lactamase gene carried on the *Escherichia coli* plasmid pBR322 so that immunologically active segments of viral gene products could be detected directly by solid-phase immunochemical methods.[14] These procedures, which are now well known, rapidly led to the expression of HBcAg in *E. coli*[15] and to the generation of a series of cloned HBV DNA fragments from which the nucleotide sequence of the viral genome was determined[16, 17] and which showed the genetic organization of the virus (Fig. 2). Although the initial yield of HBcAg was not high, it was shown to be highly immunogenic[16] and was found to be particulate and when examined in the electron microscope after immunoprecipitation (Fig. 3(a)) it was morphologically virtually identical with HBcAg isolated from the liver of an acutely infected patient.[18] Further manipulation of the cloned coding sequence for HBcAg gave much higher yields of the antigen in *E. coli* (Fig. 3(b))[19] and the product has been successfully developed for diagnostic kits that are now very widely used for detection of antibodies to HBcAg, which not only reflect infection with HBV but also serve as a surrogate marker for hepatitis non-A non-B.[20]

HBsAg AND VACCINATION AGAINST HBV

Although HBsAg expression was observed in *E. coli* by following the procedures used for HBcAg expression,[15] yields were very poor; the material was immunogenic,[21] but in spite of considerable effort in several laboratories effective systems for production of HBsAg in *E. coli* have yet to be made. However, the detailed information from the DNA sequence of HBV enabled the coding sequence for HBsAg to be manipulated in expression vectors that function in the yeast *Saccharomyces cerevisiae*, an organism widely used in large scale fermentation. HBsAg was successfully expressed in yeast strains[22–24] and although heterogeneous in cell extracts the product assembles to form particulate structures (Fig. 3(c)) closely resembling the 22 nm particles of HBsAg found naturally in the plasma of infected individuals. Such preparations proved to be immunogenic and protected chimpanzees against challenge with a high dose of the virus.[24] HBsAg is now in routine production from transformed yeast strains on a large scale, and following extensive trials in human volunteers[25, 26] preparations have now been officially approved for routine use as a vaccine in humans; several accounts of clinical aspects of these studies are to be found in Ref. 10.

In the context of this volume it is perhaps appropriate to note that the cloning and expression of HBV genes in microorganisms has not only been of scientific interest and social benefit, but also (like α-interferon) a

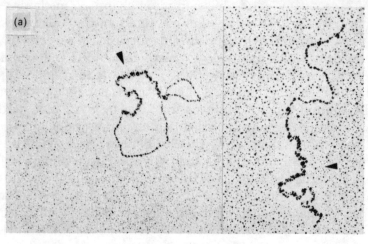

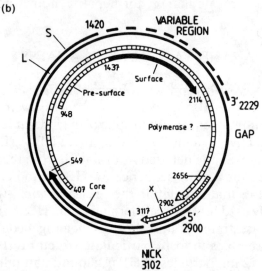

Fig. 2. The HBV genome. (a) Electron micrographs of DNA extracted from Dane particles and complexed with bacteriophage T4 gene 32 protein which binds specifically to single-stranded DNA, indicated by the arrowhead. Left, intact molecules, and right, DNA preparations digested with restriction and endonuclease *Eco*RI (for which this particular DNA contained one target). These photographs were kindly provided by Dr Hajo Delius. (b) The genome of HBV. Heavy lines denote the DNA strands, the broken line showing the region of variable length of the short strand. Arrows represent the four open reading frames (as coding sequences) with the numbers of initiation and termination triplets in the system adopted by Pasek *et al.*[16]

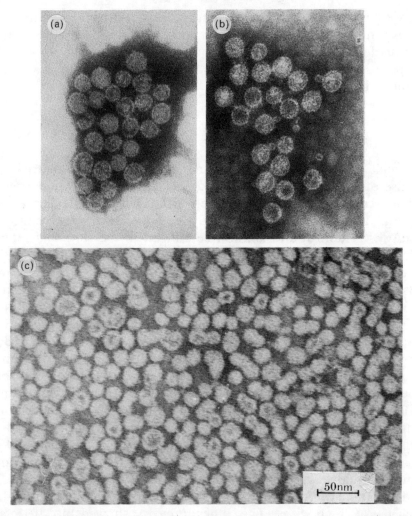

Fig. 3. Particulate forms of HBcAg: (a) isolated from the liver of an acutely infected patient; (b) synthesized in *E. coli*. These electron micrographs were kindly supplied by Dr Joan Richmond and are reprinted with permission from *Nature,* **296** (1987) pp. 667–8. Copyright 1987 Macmillan Magazines Ltd. (c) HBsAg particles purified from genetically engineered yeast. The electron micrograph was kindly provided by Dr H. Arimura and Dr T. Suyama, Green Cross Corporation, Osaka.

significant commercial success. The world wide markets for diagnostic reagents for detection of viral hepatitis and vaccines against HBV at present are almost US $500 million annually, and in a recent press release Mr James Vincent, the chief executive officer of Biogen Inc. (the sponsor of our research on HBV) stated that 'Biogen's hepatitis B licensing program, which includes both vaccine and diagnostic products, will yield a base $10 million per year in product-driven revenues which will expand as the use of recombinant products increases'. Although it is regrettable that this project has not been of direct value to a British company, it is gratifying that the University of Edinburgh will be a financial beneficiary of it.

HBcAg AND ITS ROLE IN IMMUNIZATION

Although the development of vaccines against HBV has centred upon the use of HBsAg, other immunological determinants are carried on the surface of Dane particles, namely HBeAg, which is a derivative of HBcAg, and the relatively recently demonstrated pre-S epitopes.[27] HBcAg is seldom found as the free antigen in the serum of infected individuals, but it is manifest as HBeAg, and antibodies of HBcAg appear very early after infection and before those against HBeAg or HBsAg. Further, HBcAg rather than HBsAg is the antigen found on the surface of hepatocytes in patients with chronic infections and is the antigen to which lymphocytes are sensitized, suggesting a role for HBcAg in cell-mediated immunity.[28] Given the ready availability of HBcAg through genetic engineering methods, the potential of this antigen for immunization could be assessed directly even though this had been generally discounted because the antigen is buried within the virus.

Chimpanzees were therefore immunized with HBcAg on aluminium hydroxide adjuvant and challenged with a high intravenous dose of the virus following the anticipated serum conversion.[24, 29] The results of two experiments are summarized in Fig. 4. In the first, one of the animals (Rinka) developed a short-lived and low-titre HBsAg antigenaemia, but no detectable HBeAg, and rapidly produced antibodies to both of these antigens at high titre. The other (Gwen) showed no detectable antigenaemia and at first appeared to have been completely protected against HBV, but about 5 months after the challenge the anti-HBe titre began to increase and anti-HBs was also produced showing that the animal had, in fact, suffered a very mild, serologically undetectable infection. The HBcAg used in this experiment had some HBeAg reactivity and the animal (Gwen) that had been essentially immune to the challenge inoculation had in fact developed a low anti-HBe titre as a result of the immunization with HBcAg, whereas her partner (Rinka) had not and showed the early evidence of a mild infection (Fig. 4). In the second experiment the HBcAg was again given on aluminium hydroxide adjuvant, but was first treated with sodium dodecyl sulphate to generate an enhanced HBeAg reactivity and one of the animals (Brigitte) was then

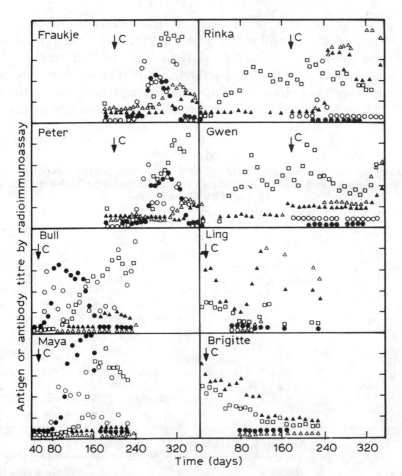

Fig. 4. Serological analysis on four chimpanzees immunized with HBcAg (right-hand panels) and four unvaccinated control animals (left-hand panels). □, Anti-HBc; O, HBsAg; △, anti-HBs; ●, HBeAg; ▲, anti-HBe. See also Table 1.

found to be completely protected against the challenge infection. The second animal (Ling), however, again appeared to have suffered a very mild infection, for a high level of anti-HBs and an enhanced titre of anti-HBe developed very rapidly after inoculation with the virus, even though there had been no HBeAg and only extremely low levels of HBsAg for a very short period (Fig. 4).

These results, together with similar observations by Iwarson *et al.*[30] with the same antigen preparations, were taken to indicate a substantial degree of protection against infection.[24, 29] The expression in *E. coli* of two other HBV genes, those for the X-antigen[17] and the DNA polymerase of the virus[31] provided reagents for the detection of antibodies to these two viral products in the sera of infected individuals or animals. The appearance of these antibodies is an additional indicator of viral replication following infection, and analysis of the sera of both immunized and control animals for these markers before and after inoculation with the virus shows that they were present only in the four control animals that exhibited the normal pattern of infection (Table 1).

Table 1

The Occurrence of HBV Antigens and Antibodies to the Antigens in Chimpanzee Serum Samples before (b) and after (a) Intravenous Inoculation with a Challenge Dose ($10^{3.5}$ ID_{50}) of HBV Serotype *ayw*

	Unvaccinated (control) animals							
Antigen or antibody	Chimpanzee							
	Fraukje		Peter		Bull		Maya	
	b	a	b	a	b	a	b	a
HBsAg	−	+	−	+	−	+	−	+
HBeAg	−	+	−	+	−	+	−	+
Anti-HBc	−	+	−	+	−	+	−	+
Anti-HBs	−	+	−	+	−	+	−	+
Anti-HBe	−	+	−	+	−	+	−	+
Anti-HBx	−	+	−	+	−	+	−	+
Anti-HBp	−	+	−	+	−	+	−	−

	Animals vaccinated with HBV antigens											
Antigen or antibody	HBsAg (yeast)				HBcAg (E. coli)				HBcAg + HBeAg (E. coli)			
	Dolf		Ianthe		Gwen		Rinka		Brigitte		Ling	
	b	a	b	a	b	a	b	a	b	a	b	a
HBsAg	−	−	−	−	−	−	−	+	−	−	−	(±)
HBeAg	−	−	−	−	−	−	−	−	−	−	−	−
Anti-HBc	−	−	−	−	+	+	+	+	+	+	+	+
Anti-HBs	+	+[a]	+	+[a]	−	+[b]	−	+[c]	−	−	−	+[c]
Anti-HBe	−	−	−	−	+	+[b]	−	+[c]	+	+[a]	+	+[c]
Anti-HBx	−	−	−	−	−	−	−	−	−	−	−	−
Anti-HBp	−	−	−	−	−	−	−	−	−	−	−	−

[a] No observable increase in antibody titre at any time after challenge with the virus.
[b] Antibody appeared or increased in titre, but not until some 5 months after challenge with the virus.
[c] Very early appearance of high-titre antibody or rapid and early elevation of antibody titre after challenge.

This would suggest that in the four animals immunized with HBcAg there was little or no replication of the virus administered as the challenge, even in the two (Rinka and Ling) that generated high anti-HBs and anti-HBe responses.[32]

It is particularly interesting to examine these results in the light of the recent observations of Milich *et al.*[33, 34] on the immune response of inbred strains of mice to HBcAg and to synthetic oligopeptides contained within the amino acid sequences of HBcAg. A single inoculation of HBcAg in Freund's complete adjuvant induced quite different antibody responses in various strains of mice and so a series of ten synthetic oligopeptides (Fig. 5) was used to search for sequences that would stimulate T-cell proliferation in groups of mice of defined H2 haplotype which had been primed with HBcAg. Two of the peptides (heavy underlining in Fig. 5) were similar to HBcAg itself in their effect on both antibody production *in vivo* and T-cell proliferation *in vitro* in specific strains of mice and some of the other peptides elicited these effects to a lesser extent in other strains. Further, it was found that these particular peptides could substitute for HBcAg in priming for the proliferation of T cells on subsequent stimulation with HBcAg or the peptide. With two strains, B10.S and Balb/c, the analysis was extended by using HBV (which cannot replicate in the mouse) for challenging after priming with HBcAg or peptides 85–100 or 120–140, and this led to the production of antibodies to HBsAg (including pre-S epitopes) as well as anti-HBc. However, when HBV was replaced as the challenge inoculation by a mixture of HBcAg and HBsAg particles, no antibodies were generated

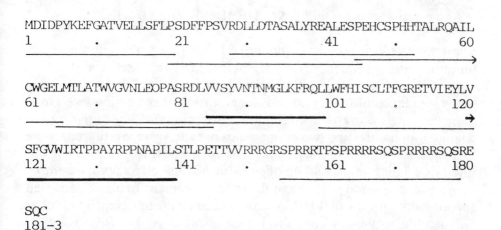

```
MDIDPYKEFGATVELLSFLPSDFFPSVRDLLDTASALYREALESPEHCSPHHTALRQAIL
1              21              .41              .              60

CWGELMTLATWVGVNLEOPASRDLVVSYVNTNMGLKFRQLLWFHISCLTFGRETVIEYLV
61             81               101             .              120

SFGVWIRTPPAYRPPNAPILSTLPETTVVRRRGRSPRRRTPSPRRRRSQSPRRRRSQSRE
121            141              161             .              180

SQC
181-3
```

Fig. 5. The amino acid sequence of HBcAg in the single-letter code. Sequences underlined are those of the synthetic peptides used by Milich *et al.*,[34] heavy underlining indicates those most active in T-cell priming.

against any of the HBsAg or pre-S epitopes, although anti-HBc was produced as usual,[34] thus demonstrating the necessity for physical linkage of the challenging epitope to the molecule or particle used for priming.

These exciting results obtained in mice clearly have significance for vaccine development if they can be extrapolated to man, for a range of immunological targets can be physically linked to HBcAg and this antigen can be produced readily in *E. coli*.[19] The results of our experiments with chimpanzees immunized with HBcAg described above (Fig. 4) certainly encourage this view. One of the four animals was completely protected and another suffered an infection too mild for detection, but which left the characteristic imprint of HBV upon the animal's immune system in that anti-HBs and anti-HBe appeared at the normal time after infection. The other two animals (Rinka and Ling) showed early and strong responses in anti-HBs production and anti-HBe elevation which it is attractive to attribute to priming of T-helper cells by the HBcAg inoculations, by analogy with the observations of Milich *et al.*[34] in mice. None of the four animals developed antibodies to the X antigen or polymerase, which may be regarded (like HBsAg) as markers of viral replication and which were produced by the control animals (Table 1). That only two of the four immunized animals responded in this way may reflect differences in histocompatibility types among outbred animals and the use of less potent adjuvant, aluminium hydroxide.

GENE FUSIONS AND DERIVATIVES OF HBcAg

Expressions of the coding sequence of HBcAg in *E. coli* gives a highly immunogenic product[16] that assembles to form particles that are virtually indistinguishable from natural core particles.[18] In manipulating the original recombinant plasmids for elevation of expression levels in *E. coli*, sequences corresponding to the *N*-terminus of the antigen were changed so that the first two amino acids of HBcAg were replaced by the first eight residues of *E. coli* β-galactosidase and a linker sequence, Glu-Phe-His.[19] This change had no observable adverse effect upon assembly of the molecules into particles or the immunogenicity of particles, which encouraged the view that HBcAg may enhance the antigenicity of other polypeptide sequences linked to it as a result of genetic fusions.

Our early attempts to fuse the coding sequence for HBsAg in its entirety or in halves at or near the 3′ end of the coding sequence for HBcAg resulted in the expression of products that displayed HBcAg antigenicity in radioimmunoassays, but showed no cross reactivity with anti-HBs in either the Abbott 'Ausria' test or in double antibody radioimmunoprecipitation assays (S. Stahl and K. Murray, unpublished work). Following the demonstration that HBcAg synthesized in *E. coli* could be converted into HBeAg,[12] the products of a series of deletions from the HBcAg coding sequence were examined with respect to their antigenic, immunogenic and biophysical characteristics (S. Stahl, P. Wingfield, B. Fearns, R. S. Tedder and K. Murray, unpublished work). The protamine-like carboxy-terminal region of the antigen could be removed without impairing assembly of particles and a number of gene fusions were therefore made that would yield a range of different epitopes fused to the carboxy-terminal region of a truncated form of HBcAg; some of these are described as their translation products in Fig. 6.

All of the polypeptides shown in Fig. 6 were produced in reasonably good yield in *E. coli* and formed particles that simplified their recovery from cell extracts. They exhibited strong antigenic activity in their cross reactions with anti-HBc and anti-HBe, and were good immunogens. In addition to high titres of anti-HBc, sera from immunized rabbits were shown to have high titres of antibody to the epitope contained in the added carboxy-terminal region in the case of the pre-S1 and pre-S2 constructions. The fusions of the HBsAg segments to HBcAg were particularly interesting; they showed no antigenic behaviour in standard radioimmune or ELISA tests for HBsAg, but they nevertheless elicited an antibody response in rabbits (reminiscent of the products of direct expression of HBsAg sequences in *E. coli*[21]), giving sera that displayed high titres of antibodies against HBsAg either on solid phase or in solution.[57]

The HBcAg derivatives carrying the short sequences from the envelope protein of human immunodeficiency virus (HIV) were examined for their immunogenicity in Balb/c mice and one of them (HBcAg-E46) stimulated production of antibodies against both HBcAg and the HIV peptide. The sequence of HIV envelope carried by this fusion protein is 728–751; antibodies raised against a synthetic peptide comprising amino acid residues 735–752 have been found to precipitate the native gp160 protein of HIV-1[35] and to neutralize the virus.[36, 37]

The results of these experiments, and a somewhat similar one by Clarke *et al.*[38] in which a peptide corresponding to amino acid residues 137–162 of the coat protein of foot and mouth disease virus (FMDV) was fused to the amino terminus of HBcAg via an appropriate genetic construction, give credence to the anticipation that fusion of antigenic peptide sequences to HBcAg is likely to constitute a general method for potentiation of the immune response to the additional antigenic or

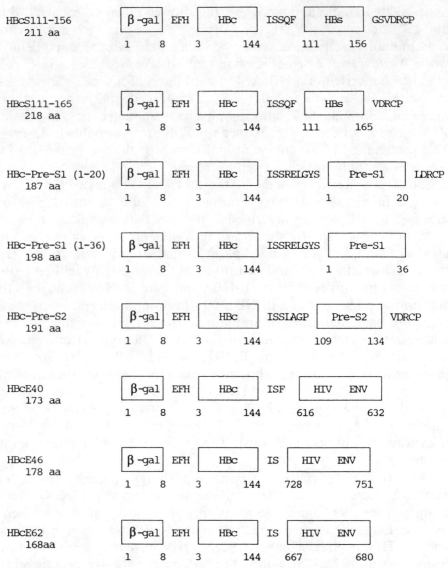

Fig. 6. Predicted amino acid sequences of HBcAg fusion proteins. The names and total amino acids of the proteins are on the left. The numbers below the boxes indicate the positions of the amino acids in the original protein. HIV ENV, HIV envelope protein (GP160).

immunogenic site. Where this also enhances priming of T-cell memory, the fusion proteins should have potential for development of vaccines and offer the prospect of producing single molecules that serve as an effective multivalent vaccine against HBV and other infectious agents.

ALTERATION OF IMMUNOLOGICAL SPECIFICITY BY MUTATION

There are a number of serological variants of HBsAg. Most isolates display the dominant *a* epitope and two sets of subtype epitopes described as *d* or *y*[39] and *w* or *r*,[40] which are believed to be to a large extent mutually exclusive so that the four serotypes generally encountered are *adw*, *adr*, *ayw* and *ayr*. Most of the antibodies induced by HBsAg are against the *a* epitope, and subtypic antisera are said to contribute little to protective immunity.[41, 42]

Several publications describe the use of synthetic peptides representing predicted antigenic regions of the HBsAg sequence to generate antisera which were then used in serological reactions with HBsAg of defined serotype. These experiments, which are summarized in Fig. 7, indicate that the dominant *a* epitope lies between residues 138 and 149 and that the region between residues 110 and 139 is involved in specification of the *d* or *y* subtype. However, modification of the conformation of HBsAg by reduction or treatment with detergent greatly reduces its immunoreactivity, suggesting that distinct peptide domains collectively constitute discontinuous epitopes.[43-46] Furthermore, some *y*-specific and *d*-specific monoclonal antibodies fail to cross react with these oligopeptides[47] indicating that *d*- and *y*-antigenicity may, like *a*-antigenicity[48] be governed by more than one epitope.

To identify residues that are critical for antigenicity and to define the *a*, *y* and *d* epitopes more precisely we have made a number of mutants in

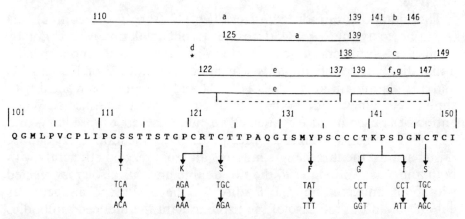

Fig. 7. The major antigenic region of HBsAg. Lines above the sequence denote synthetic peptides used to locate the *a* epitope region: a, Gerin *et al.*;[50] b, Hopp and Woods;[51] c, Prince *et al.*;[52] e, Dreesman *et al.*;[53] f, g, Bhatnaga *et al.*,[54] and Brown *et al.*[55] d is a point mutation with little effect upon *a* reactivity.[56] The amino acid substitutions and the mutations used to make them are indicated below the sequence.

HBsAg at residues that we believed, on the basis of a comparison of amino acid sequences of HBsAg of different serotypes, to correlate with the display of the appropriate antigenic subtype. These mutants are defined in Fig. 7; four of them, at residues 124 (cysteine), 142 (proline) and 147 (cysteine) were made for studies of the *a* epitope, and the others, involving residues 113 (serine), 122 (arginine) and 134 (tyrosine), were made to explore the *y* versus *d* subtype determination.

Titration of the wild-type and mutant HBsAg preparations (obtained from appropriately transformed yeast strains) with anti-HBs showed that replacement of cysteine 124 or 147 by serine drastically reduced cross reactivity with the antibodies, with the change at 147 virtually eliminating it. The proline at 142 was shown to be essential for full antigenicity, but replacement by glycine impaired cross reactivity with antibody to a lesser extent than replacement by isoleucine. All four changes also influenced (reduced) the antigen's ability to interact with a *y*-monospecific antibody and in addition virtually eliminated cross reactivity with a subtype-specific monoclonal antibody.[58, 59] These results emphasize the immunological importance of this region of the HBsAg polypeptide and reinforce the view that the epitopes are conformational.

The HBsAg used in these studies, defined as wild type, was derived from a coding sequence cloned from a virus isolated from serum exhibiting the complex serotype *adyw*,[15] but in antibody precipitation reactions the antigen showed a clear specificity for a monospecific anti-*y* serum and no interaction with a monospecific anti-*d* serum. It also induced antibodies (in mice) that cross reacted only with HBsAg of the *y* sub-type.

The mutations introduced at residues 113 (serine to theonine), 124 (arginine to lysine) or 134 (tyrosine to phenylalanine) were not by themselves sufficient to change the interaction with monospecific antibodies, but double mutants at residues 113 and 122 or 113 and 134 and a triple mutant at residues 113, 122 and 134 all cross reacted with similar efficiency with both *y*- and *d*-monospecific antisera. When antibodies raised in mice against all six mutants were exposed to *y*- and *d*-specific antigens in immunoprecipitation competition assays, those raised against the three single mutants all cross reacted efficiently with both antigens. Antibodies to the two double mutants also cross reacted with both antigens, but the avidity for the *d* subtype antigen was enhanced, and in the case of the triple mutant the induced antibodies showed no reaction with the *y* subtype antigen, but reacted very strongly with the *d* antigen.

These results, which are summarized in Table 2, show that even without determination of the detailed structure of an antigen, site-

Table 2

Antigenic and Immunogenic Characteristics of HBsAg and its Mutants[a]

Antigen or mutant	Cross reactivity with monospecific antibody[b] subtype ($\mu g\,ml^{-1}$)		Cross reactivity of induced antibody[c] subtype	
	y	d	y	d
Wild type	0·21	No inhibition	1 : 400	No inhibition
y-113	0·22	No inhibition	1 : 125	1 : 11
y-122	0·23	No inhibition	1 : 22	1 : 22
y-134	0·22	No inhibition	1 : 100	1 : 40
y-113, 122	0·23	0·23	1 : 32	1 : 562
y-113, 134	0·22	0·27	1 : 32	1 : 40
y-113, 122, 134	0·26	0·25	No reaction	>1 : 1000

[a] The results are expressed as interpolated values from titration experiments in which the antigen or antibody under test competed with its counterpart in double-antibody radioimmuno-precipitation (DARIP) assays with ^{125}I-labelled HBsAg of y or d subtype and monospecific antisera of the same subtype.
[b] Concentration of HBsAg giving 50% maximum inhibition of subtype-specific antigen–antibody interaction.
[c] Serum dilution at 50% maximum subtype-specific antigen capture.

directed mutagenesis can be used to change the immunological specificity of a protein. The examples described here are largely of academic interest, but a more practical application might involve mutant antigens that induce antibodies with higher affinities for a virus, or with a different or wider specificity to provide immunity against antigenic variants. The results also underscore the importance of precise conformation in dictating the selectivity and specificity of immunological reactions, and they further emphasize the distinction between antigenic and immunogenic epitopes.

CONCLUSION

The work described on hepatitis B provides an example of the successful application of modern biotechnological approaches to the development of an effective and commercially viable vaccine against a serious viral disease that constitutes a major public health problem worldwide. Similar approaches are in principle applicable to other vaccines; for example, immunization against rabies with a viral glycoprotein expressed in a recombinant vaccinia virus was shown to protect rabbits and mice against the virus[49] and much effort has been directed to the development of a subunit vaccine against FMDV. Work is proceeding in many laboratories on a number of viral, bacterial and parasitic diseases, much of which was described at a Nobel Symposium in 1987 and published in *Vaccine*[60] and there is, of course, widespread activity in the very difficult quest for vaccines against HIV.

ACKNOWLEDGEMENTS

Much of the work described in this chapter was supported by Biogen Inc. One of us (P.G. A.-R.) thanks the Science and Engineering Research Council and the Roseanne Campbell Hepatitis Research Fund for studentships. This article is reproduced with the kind permission of the Royal Society, and with minor modifications, from *Phil. Trans. R. Soc. Lond., B,* **324** (1989) 461–76.

REFERENCES

1. Szmuness, W., Stevens, C. E., Harley, E. J., Zang, E. A., Oleszko, W. R., William, D. C. *et al.*, Hepatitis B vaccine. Demonstration of efficacy in a controlled clinical trial in a high-risk population in the United States. *New Engl. J. Med.,* **203** (1980) 833–41.
2. Murray, K., Genetic engineering: possibilities and prospects for its application in industrial microbiology. *Phil. Trans. R. Soc. Lond.,* **B 290** (1980) 369–86.
3. Evans, D. M. A., Dunn, G., Minor, P. D., Schild, G. C., Cann, A. J., Stanway, G., Almond, J. W., Currey, K. & Maizel, J. V. Jr, Increased neurovirulence associated with a single nucleotide change in a monocoding region of the Sabin type 3 poliovaccine genome. *Nature, Lond.,* **314** (1985) 548–50.
4. Dane, D. S., Cameron, C. H. & Briggs, M., Virus-like particles in serum of patients with Australia-antigen-associated hepatitis. *Lancet,* i (1970) 695–8.
5. Vyas, G. N., Cohen, S. N. & Schmid, R. (eds), *Viral Hepatitis.* Franklin Institute Press, Philadelphia, 1978.
6. Szmuness, W., Alter, H. J. & Maynard, J. E. (eds), *Viral Hepatitis 1981. International Symposium.* Franklin Institute Press, Philadelphia, 1982.
7. Vyas, G. N., Dienstag, J. L. & Hoofnagle, J. H. (eds), *Viral Hepatitis and Liver Disease.* Grune & Stratton, Orlando, Florida, 1984.
8. Murray, K., The Leeuwenhoek Lecture, 1985. A molecular biologist's view of viral hepatitis. *Proc. R. Soc. Lond.,* **B 230** (1987) 107–46.
9. Ganem, D. & Varmus, H. E., The molecular biology of the hepatitis B viruses. *Ann. Rev. Biochem.,* **56** (1987) 651–93.
10. Zuckerman, A. J. (ed.), *Viral Hepatitis and Liver Disease.* Alan R. Liss, New York, 1988.
11. Heermann, K. H., Goldmann, U., Schwartz, W., Seyffarth, T., Baumgarten, H. & Gerlich, W. H., Large surface proteins of hepatitis B virus containing the

pre-S sequence. *J. Virol.,* **52** (1984) 396–402.

12. Mackay, P., Lees, J. & Murray, K., The conversion of hepatitis B core antigen synthesised in *E. coli* into e antigen. *J. Med. Virol.,* **8** (1981) 237–41.

13. Sureau, C., Romet-Lemmone, J. L., Millins, J. I. & Essex, M., Production of hepatitis B virus by a differentiated human hepatoma cell line after transformation with cloned circular HBV DNA. *Cell,* **47** (1986) 37–47.

14. Broome, S. & Gilbert, W., Immunological screening method to detect specific translation products. *Proc. Nat. Acad. Sci., USA,* **75** (1978) 2746–9.

15. Burrell, C. J., MacKay, P., Greenaway, P. J., Hofschneider, P.-H. & Murray, K., Expression in *Escherichia coli* of hepatitis B virus DNA sequences cloned in plasmid pBR322. *Nature, Lond.,* **279** (1979) 43–7.

16. Pasek, M., Goto, T., Gilbert, W., Zink, D., Schaller, H., MacKay, P., Leadbetter, G. & Murray, K., Hepatitis B virus genes and their expression in *Escherichia coli. Nature, Lond.,* **282** (1979) 575–9.

17. Pugh, J. C., Weber, C., Houston, H. & Murray, K., Expression of the X gene of hepatitis B virus. *J. Med. Virol.,* **20** (1986) 229–46.

18. Cohen, B. J. & Richmond, J. E., Electron microscopy of hepatitis B core antigen synthesised in *E. coli. Nature, Lond.,* **296** (1987) 667–8.

19. Stahl, S., MacKay, P., Magazin, M., Bruce, S. A. & Murray, K., Hepatitis B virus core antigen: its synthesis in *Escherichia coli* and application in diagnosis. *Proc. Nat. Acad. Sci., USA,* **79** (1982) 1606–10.

20. Koziol, D. E., Holland, P. V., Alling, D. W., Melpolder, J. C., Solomon, R. E., Purcell, R. H., Hudson, L. M., Shoup, F. J., Krakauer, H. & Alter, H. J., Antibody to hepatitis B core antigen as a paradoxical marker for non-A, non-B hepatitis agents in donated blood. *Ann. Int. Med.,* **104** (1986) 488–95.

21. MacKay, P., Pasek, M., Magazin, M., Kovacic, R. T., Allet, B., Stahl, S. *et al.,* Production of immunologically active surface antigens of hepatitis B virus by *Escherichia coli. Proc. Nat. Acad. Sci., USA,* **78** (1981) 4510–14.

22. Valenzuela, P., Medina, A., Rutter, W. J., Ammerer, G. & Hall, B. D., Synthesis and assembly of hepatitis B virus surface antigen particles in yeast. *Nature, Lond.,* **298** (1982) 347–50.

23. Miyanohara, A., Toh, E.-A., Nozaki, C., Hamada, F., Ohtomo, N. & Matsubara, K., Expression of hepatitis B surface antigen in yeast. *Proc. Nat. Acad. Sci., USA,* **80** (1983) 1–5.

24. Murray, K., Bruce, S. A., Hinnen, A., Wingfield, P., van erd, P. M. C. A., de Reus, A. & Schellekens, H., Hepatitis B virus antigens made in microbial cells immunise against viral infection. *EMBO J.,* **3** (1984) 645–50.

25. Davidson, M. & Krugman, S., Immunogenicity of recombinant yeast hepatitis B vaccine. *Lancet,* **i** (1985) 108–9.

26. Ichida, F., Yoshikawa, A., Mizokami, M., Yamamoto, M., Inaba, N., Takamizawa, H., Ohmura, T., Ohmizu, A., Ohata, J., Uemura, Y. & Nishida, M., Clinical study of recombinant hepatitis B vaccine. *J. Int. Med. Res.,* **16** (1988) 231–6.

27. Neurath, A. R., Kent, S. B. H., Strick, N., Taylor, P. & Stevens, C. E., Hepatitis B virus contains pre-S gene-encoded domains. *Nature, Lond.,* **315** (1985) 154–6.

28. Vento, S., Hegarty, J. E., Alberti, A., O'Brien, C. J., Alexander, G. J. M., Eddleston, A. L. W. F. & Williams, R., T lymphocyte sensitization to HBcAg and T cell-mediated unresponsiveness to HBsAg in hepatitis B virus-related chronic liver disease. *Hepatology,* **5** (1983) 192–7.

29. Murray, K., Bruce, S. A., Wingfield, P., van Eerd, P., de Reus, A. & Schellekens, H., Protective immunisation against hepatitis B with an internal antigen of the virus. *J. Med. Virol.*, **23** (1987) 101–7.
30. Iwarson, S., Tabor, E., Thomas, H. C., Snoy, P. & Gerety, R. J., Protection against hepatitis B virus infection by immunization with hepatitis B core antigen. *Gastroenterology*, **88** (1985) 763–7.
31. McGlynn, E. & Murray, K., Hepatitis B virus polymerase: expression of its gene in *Escherichia coli* and detection of antibodies to the product in convalescent sera. In *Viral Hepatitis and Liver Disease*, ed. A. J. Zuckerman. Alan R. Liss, New York, 1988, pp. 323–9.
32. Murray, K., Application of recombinant DNA techniques in the development of viral vaccines. *Vaccine*, **6** (1988) 164–74.
33. Milich, D. R., McLachlan, A., Moriarty, A. & Thornton, G. B., Immune response to hepatitis B virus core antigen (HBcAg): localization of T cell recognition sites within HBcAg/HBeAg. *J. Immunol.*, **139** (1987) 1223–31.
34. Milich, D. R., McLachlan, A., Thornton, G. B. & Hughes, J. L., Antibody production to the nucleocapsid and envelope of the hepatitis B virus primed by a single synthetic T cell site. *Nature, Lond.*, **329** (1987) 547–9.
35. Kennedy, R. C., Kenkel, R. D., Pauletti, D., Allan, J. S., Lee, T. H., Essex, M. & Dreesman, G. R., Antiserum to a synthetic peptide recognizes the HTLV-III envelope glycoprotein. *Science, Wash.*, **231** (1986) 1556–9.
36. Chanh, T. C., Dreesman, G. R., Kanda, P., Linette, G. P., Sparrow, J. T., Ho, D. D. & Kennedy, R. C., Induction of anti-HIV neutralizing antibodies by synthetic peptides. *EMBO J.*, **5** (1986) 3065–71.
37. Dalgleish, A. G., Chanh, T. C., Kennedy, R. C., Kanda, P., Clapham, P. R. & Weiss, R. A., Neutralization of diverse HIV-1 strains by monoclonal antibodies raised against a gp41 synthetic peptide. *Virology*, **165** (1988) 209–15.
38. Clarke, B. E., Newton, S. E., Carroll, A. R., Francis, M. J. *et al.*, Greatly improved immunogenicity of a peptide epitope by fusion of hepatitis B core protein. *Nature, Lond.*, **330** (1987) 381–3.
39. Le Bouvier, G. L., The heterogeneity of Australia antigen. *J. Infect. Dis.*, **123** (1971) 671–5.
40. Bancroft, W. H., Mundan, F. K. & Russell, P. K., Detection of additional antigenic determinants of hepatitis B antigen. *J. Immunol.*, **109** (1972) 842–8.
41. Murphy, B. L., Maynard, J. E. & Le Bouvier, G. L., Viral subtypes and cross-protection in hepatitis B virus infection of chimpanzees. *Intervirology*, **3** (1974) 378–81.
42. McAuliffe, V. J., Purcell, R. H. & Gerin, J. L., Type-B hepatitis — a review of current prospects for a safe and effective vaccine. *Rev. Infect. Dis.*, **2** (1980) 470–92.
43. Benjamin, D. C., Berzofsky, J. A., East, I. J., Gurd, F. R. N., Hannum, C., Leach, S. J., Margoliash, E., Michael J. G., Miller, A., Prager, E. M., Reichlin, M., Sercarz, E. E., Smith-Gill, S. J., Todd, P. E. & Wilson, A. C., The antigenic structure of proteins: an appraisal. *Ann. Rev. Immunol.*, **2** (1984) 67–101.
44. Dreesman, G. R., Hollinger, R. B., MacCombs, R. M. & Melnick, J. L., Alteration of HBsAg determinants by reduction and alkylation. *J. Gen. Virol.*, **19** (1973) 129–34.

45. Sukeno, N., Shirachi, R., Yamaguchi, N. & Ishida, N., Reduction and reoxidation of Australia antigen: loss and reconstitution of particle structure and antigenicity. *J. Virol.,* **9** (1972) 182–3.

46. Vyas, G. N., Rao, K. R. & Ibrahim, A. B., Australia antigen (hepatitis B antigen). A conformation antigen dependent on disulphide bonds. *Science, Wash.,* **178** (1972) 1300–1.

47. Swenson, P. D., Peterson, D. L. & Hu, P. S., Antigenic analysis of HBsAg with monoclonal antibodies specific for S-protein and pre-S2 sequences. In *Viral Hepatitis and Liver Disease*, ed. A. J. Zuckerman. Alan R. Liss, New York, 1988, pp. 627–31.

48. Tedder, R. S., Yao, J. L. & Ferns, R. B., Generation of oligonucleotide-directed mutations using phosphorothioate-DNA. *Nucl. Acids Res. Commun.,* **13** (1984) 8765–85.

49. Wiktor, T. J., MacFarlan, R. I., Reagan, K. J., Dietzschold, B., Curtis, P. J., Wunner, W. H., Kieny, M.-P., Lathe, R., Lecocq, J.-P., Mackett, M., Moss, B. & Koprowski, H., Protection from rabies by a vaccinia virus recombinant containing the rabies virus glycoprotein gene. *Proc. Nat. Acad. Sci., USA,* **81** (1984) 7194–8.

50. Gerin, J. L., Alexander, H., Shih, J. W.-K., Purcell, R. H., Dapolito, G., Engle, R., Green, N., Sutcliffe, J. G., Shinnick, T. M. & Lerner, R. A., Chemically synthesised peptides of hepatitis B surface antigen duplicate the *d/y* subtype specificities and induced subtype specific antibodies in chimpanzees. *Proc. Nat. Acad. Sci., USA,* **80** (1983) 2365–9.

51. Hopp, T. P. & Woods, K. R., Prediction of protein antigenic determinants from amino acid sequences. *Proc. Nat. Acad. Sci., USA,* **78** (1981) 3824–8.

52. Prince, A. M., Ikram, H. & Hopp, T. P., Hepatitis B virus vaccine: Identification of HBsAg/*a* and HBsAg/*d* but not HBsAg/*y* subtype antigenic determinants on a synthetic immunogenic peptide. *Proc. Nat. Acad. Sci., USA,* **79** (1982) 579–82.

53. Dreesman, G. R., Sanchez, Y., Ionescu-Matiu, I., Sparrow, J. T., Six, H. R., Peterson, D. L., Hollinger, F. B. & Melnick, J. L., Antibody to HBsAg after a single inoculation of uncoupled synthetic HBsAg peptide. *Nature, Lond.,* **295** (1982) 158–60.

54. Bhatnagar, P. K., Papas, E., Blum, H. E., Milich, D. R., Nitecki, D., Karels, M. J. & Vyas, G. N., Immune response to synthetic peptide analogues of hepatitis B surface antigen for the *a* determinant. *Proc Nat. Acad. Sci., USA,* **79** (1982) 4400–4.

55. Brown, S. E., Howard, C. R., Zuckerman, A. J. & Steward, M. W., Determination of the affinity of antibodies to hepatitis B surface antigen in human sera. *J. Immunol. Methods,* **72** (1984) 41–8.

56. Antoni, B. A. & Peterson, D. L., Site-directed mutagenesis of the hepatitis B surface antigen gene: creation of a free sulphydryl group and modification of the protein in the 22 nm particle structure. In *Viral Hepatitis and Liver Disease*, ed. A. J. Zuckerman. Alan R. Liss, New York, 1988, pp. 313–17.

57. Stahl, S. & Murray, K., Immunogenicity of peptide fusions to hepatitis B virus core antigen. *Proc. Natl. Acad. Sci. USA,* **86** (1989) 6283–7.

58. Ashton-Rickardt, P. G. & Murray, K., Mutants of the hepatitis B virus surface antigen that define antigenically essential residues in the immuno-dominant *a* region. *J. Med. Virol.* (in press).

59. Ashton-Rickardt, P. G. & Murray, K., Mutations that change the immuno-
    logical subtype of hepatitis B virus surface antigen and distinguish between
    antigenic and immunogenic determination. *J. Med. Virol.* (in press).
60. Lindberg, A. A., Norby, E. & Wigzell, H., The vaccines of the future. *Vaccine,*
    6 (1988) 73–250.

Chapter 4

TOWARDS THE STRUCTURE OF MOSAIC PROTEINS: USE OF PROTEIN EXPRESSION AND NMR TECHNIQUES

MARTIN BARON, ALAN J. KINGSMAN, SUSAN M. KINGSMAN &
IAIN D. CAMPBELL*

Department of Biochemistry, University of Oxford, Oxford, UK

INTRODUCTION

Many extracellular proteins consist of numerous distinct but repeated domains. Figure 1 illustrates how relatively few domains, or modules, can be combined to make a wide variety of proteins associated with blood-clotting, fibrinolysis, complement and the extracellular matrix.[1,2] The modules are often small disulphide bonded units of about 50 amino acids. They include: epidermal growth factor-like units (G); the 'Kringle' unit (K); fibronectin 'fingers' (F1, F2 and F3); the short consensus repeat seen in the C4 binding protein and other proteins of complement (C);[3] a module which appears in thrombospondin and properdin (T);[4] one seen in the LDL receptor (A)[5] and modules with homology to the EF hands in calmodulin (E).[6] There is a close relationship between these domains and the exon/intron structure of their genes. Exon duplication and shuffling was probably involved in the evolution of these mosaic proteins. It is significant that nearly all of these modules have phase 1 exon boundaries,[7] such that exons may be duplicated, inserted or deleted without changing the reading frame for the rest of the protein. The modules have acquired diverse roles in their subsequent evolution. They appear to form autonomously folding structural units since disulphides are usually within rather than between modules. This is supported by the structures known so far.

Structural studies of mouse and human epidermal growth factors (EGF), and the related human transforming growth factor alpha (TGFα)

*To whom correspondence should be addressed.

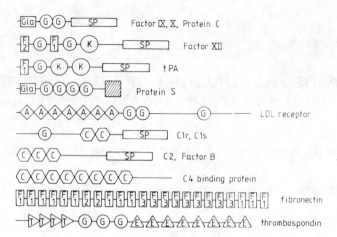

Fig. 1. Mosaic proteins from blood clotting, fibrinolytic pathways, complement and the extracellular matrix, showing their modular design. The modules are as follows: Gla, calcium binding domain containing gamma carboxylated glutamates; G, growth factor-like; K, kringle; F1, F2, F3, fibronectin type 1, 2 and 3; A, type A of the LDL receptor; C, complement short consensus repeat; E, calcium binding EF hand; T, thrombospondin repeat; SP, serine protease domain.

by nuclear magnetic resonance (NMR) reveal that the overall EGF fold is conserved although the homology is only 40%.[8] On the basis of the observed structure, Cooke *et al.*[9] made predictions about a calcium binding site on a growth-factor module of factor IX.

NMR and X-ray studies of kringle units have been carried out[10-13] and the structure of the serine protease domains and EF hands can be inferred from the known homologous structures. The structures of most of the other modules are not known.

Apart from prothrombin fragments it has proved difficult so far to obtain crystal structures for mosaic proteins. Attempts are being made to produce models based on knowledge of the structure of the modular units together with electron microscopy or small angle scattering studies of the intact protein. For this approach to be successful, structures of more module types must be determined. Information about the way the modules fit together, i.e. module/module interfaces is also important. For example, in the case of tissue plasminogen activator, structures of homologues of four of the five modules indicated in Fig. 1 are known, but there is little information about the nature of domain/domain contacts. The structure of the N-terminal F1 module is not known.

It is now possible to determine the structures of small proteins, like these modules, in solution using NMR[14-16] thus avoiding the often difficult step of crystallisation. This provided an incentive for us to produce modules and pairs of modules in sufficient quantities for NMR studies with the aim of obtaining detailed structural models of a large number of proteins of the type shown in Fig. 1.

This chapter describes our preliminary results on the production of modules for NMR studies.

EXPRESSION STRATEGY

There are three possible routes for producing modules. One is to isolate proteolytic cleavage products of larger proteins, another is to produce the domains by peptide synthesis and the third is to use a protein expression system.

Modules have previously been isolated as stable structures following proteolysis. The immunoglobulin domains are well known examples and various kringle units from plasminogen have been produced.[11] In many cases, however, there are no suitable proteolysis sites, or the intact protein itself may be difficult to obtain in sufficient quantities.

Methods of peptide synthesis have rapidly improved in recent years and EGF has been produced in this way.[17] It remains difficult, however, to produce significant amounts of peptides which are longer than about 40 residues and which contain disulphide bridges. This route is, however, becoming increasingly viable for many of the modules.

Expression systems based on *E. coli* have the advantage of being easy to set up, with many 'designer vectors' readily available. However, there are problems when expressing disulphide bonded proteins at high levels. Much of the protein aggregates into inclusion bodies. These are easy to separate and they yield fairly pure protein but it must be denatured and refolded into the correct conformation. A number of eukaryote systems are also available, including yeast, mammalian and insect cells.[18-21] All have the capability, in favourable cases, of producing sufficient quantities of correctly folded and soluble protein. Because of potential folding problems and lack of specific module assays, we chose to use a yeast secretion vector to express domains directly into the extracellular medium.

The 13 residue yeast alpha-factor mating pheromone is expressed as a 165 residue percursor.[22] The precursor consists of a hydrophobic signal sequence at the N-terminus, a 60 residue leader sequence and four copies of the alpha-factor peptide separated by spacer peptides. The spacers consist of Lys.Arg followed by a number of Glu.Ala or Asp.Ala repeats. The prepropeptide directs secretion into the culture medium, the spacers being removed by three enzymatic processing events (Fig. 2).

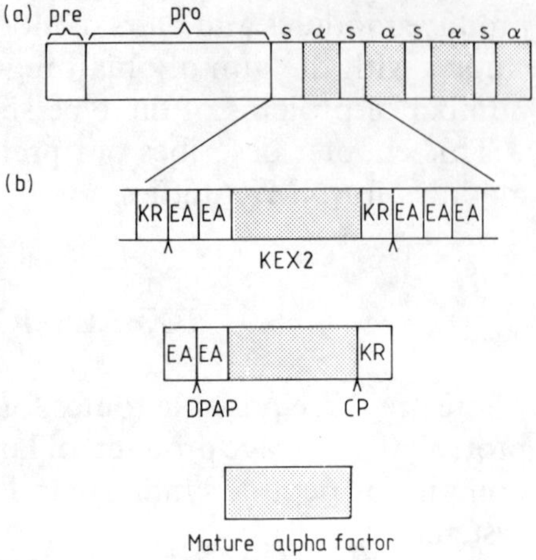

Fig. 2. (a) The precursor of the alpha-factor peptide. The pre-sequence is a 22 residue signal sequence, followed by a 60 residue propeptide. The alpha-factor peptide is repeated four times with an intervening cleavage signal sequence S. (b) An expanded view showing the three processing events leading to mature alpha-factor: *KEX2* is a cathepsin B like protease; DPAP is a dipeptidyl amino peptidase and CP is a carboxypeptidase.

A number of studies[23-25] have demonstrated that the leader sequence is capable of directing secretion into the medium of correctly folded and biologically active heterologous proteins. Brake *et al.*[23] used the leader sequence to secrete human EGF and showed that the Lys.Arg part of the spacer peptide is sufficient to obtain correct processing at the N-terminus. The removal of the Glu.Ala dipeptides was shown to be rate limiting at high expression levels and led to aberrant N-termini. Biologically active EGF was secreted into the medium at about 5 mg/litre. It was authentically processed at the N-terminus and was found to make up greater than 90% of the protein in the medium following removal of the cells. The main problem with the secretion system appears to be limited proteolysis of the product. Both internal cleavages and removal of C-terminal residues have been observed.[24, 26, 27] The extent to which this is a problem appears to vary from protein to protein.

The alpha-factor leader sequence was cloned into M13 and site-directed mutagenesis was used to introduce a HindIII site just prior to the Lys.Arg cleavage signal. No change in the amino acid coding was required. The mutated leader was cloned into the polylinker of the SP64 vector (Pronega Biotechnology) to give pMB50.

Two strategies have been used by the authors for producing the gene to be expressed: total synthesis or isolation of suitable gene fragments from available genes after cutting at suitable restriction sites. A general strategy for the construction of in-frame fusions to the leader sequence is shown in Fig. 3. The HindIII and BamHI fragment is replaced with a suitable linker for the insertion of a domain in the correct reading frame so that cleavage after the Lys.Arg releases the domain, which is secreted into the medium. The leader/domain fusion is then removed on a BglII/

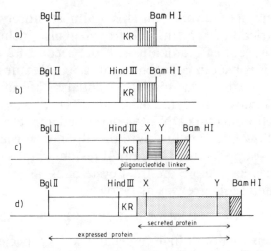

Fig. 3. (a) The alpha-factor leader peptide, showing the Lys.Arg cleavage signal. Vertical shading marks the Glu.Ala repeat which was removed during the construction. (b) Site-directed mutagenesis was used to introduce a HindIII site just prior to the Lys.Arg cleavage signal. (c) The HindIII to BamHI fragment is replaced with a suitable linker; X and Y are restriction sites into which a domain may be inserted, replacing the horizontally shaded spacer. (d) Stippled shading is the secreted domain, oblique shading has an in-frame stop codon supplied by the linker. The BglII/BamHI fragment may now be inserted into the BglII expression site of the yeast vector pMA91 (Mellor *et al.*, 1983) where transcription is driven by the yeast PGK promoter.

BamHI fragment and inserted into the BglII expression site of the yeast expression vector pMA91,[18] which uses the PGK promoter to drive transcription. Yeast cells (*Saccharomyces cerevisiae*) were transformed[28, 29] and recombinants selected by their ability to grow on a medium lacking leucine.

By using suitable linkers this strategy has been applied to a number of constructions including a fibronectin type I domain, human EGF and mutant EGFs (Dudgeon *et al.*, submitted); the EGF pair and the first EGF domain of factor IX[30] (Handford *et al.*, submitted), a factor H repeat[31] and a properdin repeat[32] (factor H and properdin are associated with the complement system). The alpha-factor secretion system is thus an attractive one for expressing domains and domain pairs, as it has proved possible to identify and purify the expressed material using amino-acid composition as an assay.

THE FIBRONECTIN TYPE 1 DOMAIN

As an example of the approach used, the results with an F1 domain from fibronectin will be discussed in some detail. The domain chosen was the seventh F1 domain of fibronectin. This is almost entirely encoded by an NcoI/NsiI restriction fragment. This fragment also includes a sequence that links the seventh F1 domain with the second F2 domain and is thought to be involved in collagen binding.[33] Our expressed domain in isolation does not however bind to a collagen affinity column. The fragment was removed from the plasmid pFH16y[33] which contains part of the human fibronectin cDNA. As described above, a suitable linker was used to allow in-frame fusion to the alpha-factor leader sequence. The linker also supplied the C-terminal residues and stop codon since the NsiI site lies just before the final cysteine.

The translated sequence was as below:

...KRPMAAHEEICTTNEGVMYRIGDQWDKQHDMGHMMRCTCVGNGRGEWTCI

With this construction, a high molecular weight species was found as the major band on SDS PAGE, suggesting that cleavage after Lys.Arg was being prevented by the proline. To overcome this the 17 base pair HindIII/NcoI sequence was replaced with a sequence which changed the proline for a serine. This led to efficient cleavage by the *KEX2* enzyme and production of the required protein. Transformed yeast cultures were grown in shake flasks for 2 days and cells removed by centrifugation. The secreted fibronectin domain, about 3 mg/litre of cells, was concentrated by batch absorption with C18 silica beads and eluted with 50% acetonitrile. After freeze drying the crude material was purified by reverse phase HPLC (Fig. 4). The major peak was collected and examined by analytical reverse phase HPLC, SDS PAGE (Fig. 5), amino acid analysis (Table 1) and N-terminal sequencing which showed that the protein had been correctly processed from the leader sequence. This method has proved quite general for a number of different domains and mutants expressed in this way.

In order to show that the expressed domain was correctly folded it was important to ascertain the disulphide bonding pattern. A combination of tryptic cleavage and a single round of manual Edman degradation were used. The latter allowed cleavage between the second and third cysteines which are only one apart in the sequence. The resulting peptides were separated by reverse phase HPLC and their sequence determined. The results indicate that the disulphide pattern is the same as for type 1 domains in the intact fibronectin molecule (data not shown).

NMR TECHNIQUES

Typical three-dimensional structure determination in solution by NMR involves the following steps.

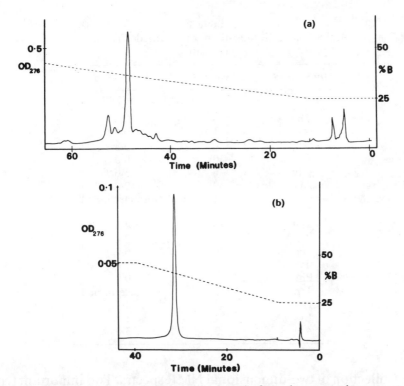

Fig. 4. Purification of the type 1 fibronectin domain from the yeast culture supernatant. (a) Reverse phase HPLC on a semi-preparative C18 column, eluting with an acetonitrile gradient (buffer A = 0·1% trifluoroacetic acid (TFA); buffer B = 80% acetonitrile, 0·1% TFA), and recording at 276 nm. The major peak was collected and analysed with an analytical C18 column (b).

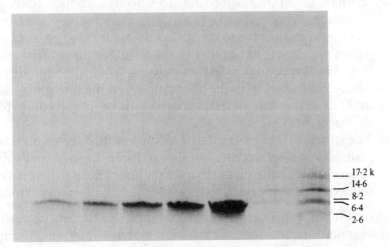

Fig. 5. SDS PAGE using a Pharmacia Phast gel with a 8–25% gradient. From the left, tracks 1–5 show increasing concentrations of purified domain; tracks 7 and 8 are Pharmacia peptide markers.

Table 1
Amino Acid Composition of Purified Fibronectin
Domain

Amino acid	Expected	Determined
Asx	5	4·8
Thr	4	3·4
Ser	1	0·9
Glx	6	5·8
Pro	0	0·0
Gly	6	5·6
Ala	2	2·0
Cys	4	Not determined
Val	2	1·7
Met	5	4·5
Ile	3	2·4
Leu	0	0·0
Tyr	1	1·0
Phe	0	0·0
Trp	2	Not determined
His	3	3·1
Lys	1	1·7
Arg	3	3·0

(1) Collection of two-dimensional NMR spectra. Two important types of information may be derived from two-dimensional experiments, one about protons less than three bonds apart (i.e. intra-residue), the other about protons which are closer than 0·5 nm apart.

(2) Through-bond and through-space information is used to assign resonances in the spectrum to specific amino acids in the sequence.

(3) Structural restraints are identified. These include pairs of protons which are close to each other in space; information on bond angles from small splittings in the resonances. In addition amide protons whose peaks persist in the spectrum when the protein is dissolved in D_2O can often be identified as those involved in forming hydrogen bonds.

(4) The structure is calculated using the identified restraints. Distance geometry methods create families of three-dimensional structures consistent with the distance information. These structures may be refined by carrying out a restrained molecular dynamics simulation.

Recombinant expression technology and NMR are powerful and complementary techniques. Expression systems may, in principle, be utilised to produce any specified protein fragment. Isotopic labelling, e.g.

with ^{15}N, simplifies the spectrum and aids the resolution of overlapping resonances, thus bringing larger proteins within reach of the NMR methodology. NMR also provides a rapid and sensitive comparison of the structures of wild type and mutant proteins which is very important in the interpretation of mutagenesis experiments. An additional benefit is that resonance assignments may be aided with site-specific changes.

Figure 6 shows a ^{1}H spectrum in H_2O of the fibronectin domain. The two alanine methyls, five methionine methyls and three pairs of histidine ring protons can be distinguished. Although they are not clearly identified in the one-dimensional spectrum, many other amino acid types have been identified by two-dimensional NMR techniques, including all three isoleucines, which demonstrates that the C-terminus

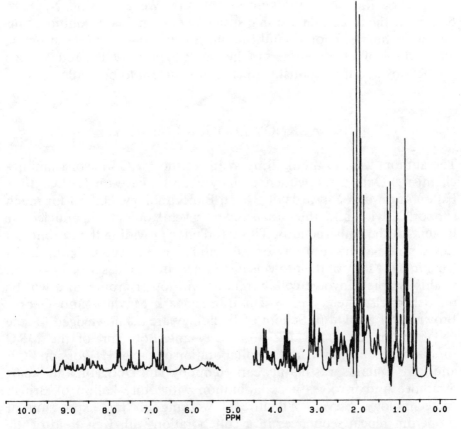

Fig. 6. A ^{1}H NMR 600 MHz spectrum of an expressed fibronectin domain. The solvent was a 9 : 1 H_2O/D_2O mixture, pH 7·6; temperature 300 K. The solvent resonance was suppressed by saturation and the residual solvent peak has been omitted for clarity.

is intact. Methyl resonances of a threonine and a valine are shifted upfield from their normal position, presumably due to the proximity of aromatic groups. The resonances between 5·0 and 5·7 ppm are characteristic of alpha protons in a beta sheet. Confirmation of this and further structural information must await complete sequential resonance assignments.

CONCLUSIONS

We have shown that milligrams of a module with the correct fold can be expressed and readily purified using an alpha factor secretion system. We have now constructed a large number of vectors with the aim of determining the structures of various modules in Fig. 1, and the way in which they interact with their neighbours. We are using NMR to determine the three-dimensional structure of domains in solution, thus avoiding the need for crystallisation. Solving one module structure allows those of other members of the family to be modelled and detailed predictions about relationships between structure to be made.

ACKNOWLEDGEMENTS

The authors wish to thank Tony Willis of the MRC unit of Immuno-chemistry, Oxford for sequence analysis and advice on HPLC; Tony Fallon and Mark Edwards of British Biotechnology, Oxford for much support, advice and the provision of several of the oligonucleotide linkers for the constructions; Tito Baralle for providing the fibronectin DNA and Sandra Fulton for her guidance on recombinant DNA techniques. Many of the projects mentioned, but not described in detail in this chapter, involve collaborations with other groups and will be published elsewhere. Penny Handford, Mark Mayhew and George Brownlee of the Dunn School of Pathology are acknowledged for the collaboration on factor IX; Tony Day and Bob Sim of the MRC Immunochemistry unit, for the collaboration on factor H. Work on EGF and EGF mutants was mainly carried out by Tim Dudgeon and Stuart Strathdee and involves a collaboration with Tony Fallon of British Biotechnology; Steven Wiltshire is working on the expression of properdin repeat sequences in a collaboration with Ken Reid of the MRC Immunochemistry unit. The authors thank the SERC and Wellcome Foundation for financial support. This is a contribution from the Oxford Centre for Molecular Science.

REFERENCES

1. Doolittle, R. F., The genealogy of some recently evolved vertebrate proteins. *Trends Biochem. Sci.,* **10** (1985) 233–7.
2. Patthy, L., Detecting distant homologies of mosaic proteins. *J. Molec. Biol.,* **202** (1988) 680–96.
3. Campbell, R. D., Law, S. K. A., Reid, K. B. M. & Sim, R. B., Structure, organization, and regulation of the complement genes. *Ann. Rev. Immunol.,* **6** (1988) 161–95.
4. Robson, K. J. H., Hall, J. R. S., Jennings, M. W., Harris, T. J. R., Marsh, K., Newbold, C. I., Tate, V. E. & Weatherall, D. J., A highly conserved amino-acid sequence in thrombospondin, properdin and in proteins from sporozoites and blood stages of a human malaria parasite. *Nature,* **335** (1988) 79–82.
5. Scott, J., Unravelling atherosclerosis: Lipoprotein receptors. *Nature,* **338** (1989) 118–19.
6. Lawler, J. & Hynes, R. O., The structure of human thrombospondin, an adhesive glycoprotein with multiple calcium binding sites and homologies with several different proteins. *J. Cell Biol.,* **103** (1986) 1635–48.
7. Patthy, L., Intron-dependant evolution: Preferred types of exons and introns *FEBS Lett.,* **214** (1987) 1–7.
8. Campbell, I. D., Cooke, R. M., Baron, M., Harvey, T. S. & Tappin, M. J., The solution structures of EGF and TGF-α. *Prog. Growth Factor Res.,* **1** (1989) (in press).
9. Cooke, R. M., Wilkinson, A. J., Baron, M., Pastore, A., Tappin, M. J., Campbell, I. D., Gregory, H. & Sheard B., The solution structure of hEGF. *Nature,* **327** (1987) 339–41.
10. Park, C. H. & Tulinsky, A., The three-dimensional structure of the kringle sequence: Structure of prothrombin fragment 1. *Biochemistry,* **25** (1986) 3977–82.
11. Tulinsky, A., Park, C. H., Mao, B. and Llinas, M., Lysine/fibrin binding sites of kringles modeled after the structure of kringle 1 of prothrombin. *Proteins,* **3** (1988) 85–96.
12. Mabbut, B. C. & Williams, R. J. P., Two dimensional ^{1}H-NMR studies of the solution structure of plasminogen kringle 4. *Eur. J. Biochem.,* **170** (1988) 539–48.
13. Harlos, K., Holland, S. K., Boys, C. W. G., Burgess, A. I., Esnouf, M. P. & Blake, C. C. F., Vitamin K-dependent blood coagulation proteins form hetero-dimers. *Nature,* **330** (1987) 82–4.
14. Wüthrich, K., *NMR of Proteins and Nucleic Acids.* John Wiley, New York, 1986.
15. Cooke, R. M. & Campbell, I. D., Protein Structure Determination by NMR. *Bioessays,* **8** (1988) 52–6.
16. Kaptein, R., Boelens, R., Scheek, R. M. & van Gunsteren, W. F., Protein structure from NMR. *Biochemistry,* **27** (1988) 5389–95.
17. Heath, W. F. & Merrifield, R. B., A synthetic approach to structure function relationships in the murine EGF molecule. *Proc. Nat. Acad. Sci., USA,* **83** (1986) 6367–71.
18. Mellor, J., Dobson, M. J., Roberts, N. A., Tuite, M. F., Emtage, J. S., White, S., Lowe, P. A., Patel, T., Kingsman, A. J. & Kingsman, S. M., Efficient synthesis of enzymatically active calf chymosin in *S. cerevisiae. Gene,* **24** (1983) 1–14.

19. Kingsman, A. J. & Kingsman, S. M., Heterologous gene expression in *Saccharomyces cerevisiae. Biotech. Genet. Engng Rev.,* **3** (1985) 377–416.
20. Rees, D. J. G., Jones, I. M., Handford, P. A., Walter, S. J., Esnouf, M. P., Smith, K. E. & Brownlee, G. G., The role of β-hydroxyaspartate and adjacent carboxylate residues in the first EGF domain of human factor IX *EMBO J.,* **7** (1988) 2053–61.
21. Smith, G. E., Summers, M. D. & Fraser, M. J., Production of human beta interferon in insect cells infected with a baculovirus expression vector. *Molec. Cell. Biol.,* **3** (1983) 2156–65.
22. Kurjan, J. & Herskowitz, I., Structure of a yeast pheromone gene (MF-α): A putative α-factor precursor contains four tandem copies of mature α-factor. *Cell,* **30** (1982) 933–43.
23. Brake, A. J., Merryweather, J. P., Coit, D. G., Heberliein, U. A., Masiarz, F. R., Mullenbach, G. T., Urdea, M. S., Valenzuela, P. & Barr, P. J., Alpha factor directed synthesis and secretion of mature foreign proteins in *S. cerevisiae. Proc. Nat. Acad. Sci., USA,* **81** (1984) 4642–6.
24. Bitter, B. A., Chen, K. K., Banks, A. R. & Por-Hsiung, L., Secretion of foreign proteins from *Saccharomyces cerevisiae* directed by α-factor gene fusions. *Proc. Nat. Acad. Sci., USA,* **81** (1984) 5330–4.
25. Mullenbach, G. T., Tabrizi, A., Blacher, R. W. & Steimer, K. S., Chemical synthesis and expression in yeast of a gene encoding connective tissue activating peptide-III. *J. Biol. Chem.,* **261** (1986) 719–22.
26. Vlasuk, G. P., Bencen, G. H., Scarborough, R. M., Tsai, P. K., Whang, J. L., Maack, T., Camargo, M. J. F., Kirsher, S. W. & Abraham, J. A., Expression and secretion of biologically active human atrial natriuretic peptide in *Saccharomyces cerevisiae. J. Biol. Chem.,* **261** (1986) 4789–96.
27. George-Nascimento, C., Gyenes, S., Halloran, S. M., Merryweather, J., Valenzuela, P., Steimer, K. S., Masiarz, F. R. & Randolph, A., Characterization of recombinant hEGF produced in yeast. *Biochemistry,* **27** (1988) 797–802.
28. Beggs, J. D., Transformation of yeast by a replicating hybrid plasmid. *Nature,* **275** (1978) 104–9.
29. Hinnen, A., Hicks, J. B. & Fink, G. R., The transformation of yeast. *Proc. Nat. Acad. Sci., USA,* **75** (1978) 1929–33.
30. Anson, D. S., Choo, K. H., Rees, D. J. G., Giannelli, F., Gould, J. A., Huddleston, J. A. & Brownlee, G. G., The gene structure of human anti-haemophilic factor IX. *EMBO J.,* **3** (1984) 1053–60.
31. Ripoche, J., Day, A. J., Harris, T. J. R. & Sim, R. B., Complete amino acid sequence of human complement factor H. *Biochem. J.,* **249** (1988) 593–602.
32. Goundis, D. & Reid, K. B. M., Properdin, the terminal complement components, thrombospondin and the circumsporozoite protein of malaria parasites contain similar sequence motifs. *Nature,* **335** (1988) 82–5.
33. Owens, R. J. & Baralle, F. E., Mapping the collagen-binding site of human fibronectin by expression in *E. coli. EMBO J.,* **5** (1986) 2825–30.

Chapter 5

PRODUCTION OF SECRETED PROTEINS IN YEAST

S. H. COLLINS

Delta Biotechnology Limited, Nottingham, UK

INTRODUCTION

Human serum albumin (HSA) is the largest single protein component of plasma[1] where its role is to maintain normal osmolarity and to act as a carrier for numerous small molecules (including nutrients and metabolites) many of which would otherwise have low solubility or be poorly tolerated in free solution. Compounds which it is capable of binding include fatty acids, bilirubin and numerous drugs.[2] Unlike many recombinant proteins currently being considered by the biotechnology industry, albumin is already sold and used in large quantities.[3] It is prepared by fractionation of donated blood and is used in the treatment of patients requiring fluid replacement.[4] It is used particularly in the treatment of burn victims, those suffering from traumatic shock and some special groups of surgical patients. However its use is much affected by the custom and practice of particular countries and the preference of individual doctors.

Current global usage is of the order of some hundreds of tonnes per annum. the US price was around $2·50/g in 1985.[3] A typical dose is dispensed in 12·5 g vials or multiples thereof.[4] Clearly any attempt to make an impact on such a market from a new recombinant source requires a multi-tonne per annum production facility. Thus in comparison with other potential biotechnology products (e.g. tPA or Factor VIII) an unusually large-scale operation to produce a relatively low cost product is being considered. These scale and cost requirements are major factors in determining the nature of a possible production process and the choice of a suitable host organism.

ORGANISM AND PROCESS CHOICE

The first consequence of the cost requirement of this product is to exclude absolutely any process involving the use of the expensive media used in mammalian cell culture so we are obliged to rely upon a recombinant micro-organism in the fermentation process. Fortunately HSA is not glycosylated[5] so differences between mammalian and microbial glycosylation pathways do not pose any difficulty. However the nature of the protein does place further restrictions on possible production processes. Albumin is a soluble protein of mol.wt 66 500 which is normally secreted from liver cells.[6] It contains 17 disulphide bonds,[7] which are a common feature of secreted proteins although the number might be regarded as unusually high for a protein of this size. It has commonly been observed that attempts to produce large quantities of such proteins in intracellular form in recombinant micro-organisms result in the accumulation of the proteins in an insoluble denatured form, often in so-called inclusion bodies, with the disulphide bonds formed incorrectly or not at all[8-10] It was found that intracellular recombinant albumin in yeast also accumulates in an insoluble form. Purification of albumin (or other recombinant proteins) so found requires cell breakage, solubilisation using denaturants (e.g. urea or guanidinium hydrochloride) and reducing agents such as glutathione or mercaptoethanol, followed by renaturation (usually by dialysis) prior to protein purification. Such a protocol has been described for recombinant albumin.[11] At an early stage in the development of the project, the use of a route for albumin production which would have required such processing was considered[12] but the resulting downstream problems render it an impossibility at the scale required. In particular the requirement for large quantities of denaturant are prohibitive in spite of improvements in the renaturation protocol.[13] The urea requirements would be similar to the output of a medium sized fertiliser plant and there is no satisfactory method of disposal of the partially diluted urea stream, which would also contain mercaptoethanol and other organic impurities, coming from the dialysis stage.

It is therefore essential to develop a process in which the albumin is produced in its native conformation. This requires albumin to be produced in such a way that it would be secreted from the recombinant micro-organism (as it is from the hepatocyte) in the hope that in following the same secretory pathway it would naturally fold up correctly. The yeast *Saccharomyces cerevisiae* was considered to be the most suitable organism for the following reasons:

(a) *E. coli* (and other Gram-negative bacteria) are able to secrete heterologous proteins[14, 15] and natural as well as bacterial leader sequences are sometimes well tolerated. However the presence of the outer membrane of the Gram-negative cell envelope[16] means that in general secreted proteins will accumulate in the periplasm. There is therefore still a need for cell disruption and the highly toxic nature of the lipopolysaccharide (endotoxin) component of the Gram-negative cell envelope[17] places severe demands on the subsequent purification.

(b) Gram-positive bacteria (e.g. Bacilli) are capable of secretion into the growth medium and recombinant HSA has indeed been produced by this route.[18] However the prokaryotic and eukaryotic secretory pathways are markedly different, which may account for the relatively inefficient secretion which was observed.[18] A further problem with such organisms is that they commonly secrete a number of proteins, which complicates the recovery and purification particularly since the commonly secreted proteins include proteases which could be expected to degrade the product.

(c) The yeast secretory pathway shares many common features with higher eukaryotes so that good secretion of albumin from yeast appeared to be a strong possibility at the outset of this project.[19] *Saccharomyces cerevisiae* was adopted as the organism of choice because of the much greater body of knowledge, particularly in molecular biology, genetics, metabolism and fermentation for this organism in comparison with any other eukaryote. *S. cerevisiae* also has the advantage that it does not normally secrete many proteins so the subsequent purification would be simplified. Furthermore the extensive use of this organism in food and beverage production is a major advantage in demonstrating that it is safe for use in fermentation processes.

THE FERMENTATION PROCESS

Let us consider how a fermentation plant designed to produce albumin in tonne quantities might operate. The amounts of albumin required imply a production capacity of hundreds or even thousands of tonnes of yeast per annum. It is essential to minimise the size and capital cost of the plant so a high plant productivity is required. This implies an intensive fermentation operating at high final biomass. Baker's yeast is currently produced at this scale using fed batch fermentation (the Zulauf process) operating at final cell concentrations as high as 100 g/litre dry weight.[20] There are, however, a number of differences between a process designed for recombinant albumin production rather than biomass. Firstly the range of conditions which can be tolerated is restricted by the stability of the recombinant protein — this particularly effects the control of pH and foam. Secondly baker's yeast fermentation uses molasses as the cheapest available carbohydrate source.[20] This is unacceptable for production of secreted recombinant proteins because the benefit of the cheap carbon source is negated by the greatly increased difficulty of subsequent purification resulting from the use of crude raw materials. We have therefore used a defined minimal medium adapted from medium Dw of Fiechter *et al.*[21] This is a minimal salts and trace elements mixture containing a few pure vitamins. pH control by ammonia addition provides the bulk of the nitrogen requirement and the carbon source may be glucose, sucrose or any other substrate which is both capable of supporting growth of *S. cerevisiae* and available in reasonable purity.

As a fermentation organism *S. cerevisiae* has one well known disadvantage; in the presence of excess carbohydrate the maximum growth rate occurs by fermentation with ethanol production even in the presence of oxygen. This is commonly referred to as the Crabtree effect[24] although this maybe something of a misnomer. The effect appears to be better described as the result of limited respiratory capacity in *Saccharomyces* type yeast[25] such that the respiratory system is only capable of supporting growth rates up to some limiting value less than the maximum growth rate, rather than an effect involving inhibition of repression of respiratory enzymes which is characteristic of the Crabtree effect proper.[26] The consequence of this effect is well illustrated by reference to observations made in carbon limited continuous culture.[27] At growth rates below some limiting value, characteristic of the yeast strain, carbohydrate is oxidised, no ethanol accumulates and cell yield is high (typically about 0·5 g/g carbohydrate). Increasing the growth rate above the limiting value results in a progressively higher proportion of the carbohydrate being fermented to ethanol with consequent loss of cell yield. This change may be detected by analysis of the fermenter off-gas.[28, 29] Since ethanol formation is accompanied by CO_2 production, CO_2 evolution rises in comparison with oxygen uptake causing a rise in respiratory quotient (RQ) which is the ratio of these two variables.

Control of a continuous *Saccharomyces* fermentation is easily achieved by selecting a dilution rate below the limiting value. A similar effect may be obtained in fed batch culture by careful control of a nutrient feed pump supplying a concentrated carbohydrate solution. This can be achieved manually but requires constant attention and frequent adjustment of the nutrient feed rate if one is to approach the optimum condition of near constant growth rate. In reality the ideal is never achieved. Much better and more reproducible control can be achieved using a computer to monitor fermentation conditions, including off gas analysis and to control the feed rate. Wang *et al.*[30] have described an algorithm for computer control of a baker's yeast fermentation. The protocol is essentially a feed-forward system controlling nutrient feed with an initial set parameter for growth rate combined with a feedback loop which makes further adjustments to the feed rate in response to RQ. This algorithm has been adapted for use with a proprietary software package for fermenter control (the micro MFCS system designed by Rintekno and marketed by B. Braun, Melsungen, Germany). Our protocol also allows for initiation of feed at an appropriate moment and prevents further increases in feed rate if the maximum oxygen transfer rate achievable by the aeration and agitation system has been reached.

The operation of this protocol to control a basic fed batch fermentation is illustrated in Fig. 1. The fermenter (101 nominal working volume) initially contained 5 litres of medium including 2% sucrose. Following inoculation there was a period of exponential growth on sucrose with ethanol production ('Crabtree effect' conditions) lasting approximately 20 h. When the carbohydrate had all been converted to ethanol there was

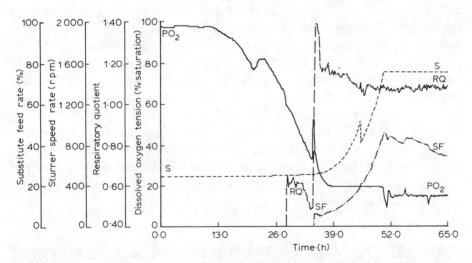

Fig. 1. Computer control profile of fed batch fermentation of recombinant yeast. SF, substrate feed rate (% of maximum value of 10 ml/min); S, stirrer speed rate; RQ, respiratory quotient (CO_2 evolution/O_2 uptake); PO_2, dissolved oxygen tension (as % air saturation at 30°C and 1 atmosphere pressure). See the text for an explanation of the sequence of events.

a small transient rise in dissolved oxygen tension (DOT) followed by a steady fall in DOT over the next 15 h or so as the ethanol was oxidised. The oxidation of the relatively reduced substrate (ethanol) is characterised by the low RQ observed in this period. (Gas analysis data are only shown for the period following the initial 28 h growth.)

When oxidation of the ethanol was complete a nutrient feed pump was started supplying a solution of 50% w/v sucrose containing additional salts and vitamins. There was an immediate rise in RQ which then settled down to a value of about 1·1 as the system came under control and near steady state exponential growth was attained. In this particular experiment the DOT controller was set to control DOT at 20% air saturation using stirrer speed control but with a minimum speed of 500 rpm. Thus the DOT fell until 20% air saturation was approached. The stirrer speed then rose gradually to accommodate the increasing oxygen uptake of the culture consequent upon the exponential increase in feed rate and biomass. The sharp reduction in agitation rate at about 45 h was caused by a step increase in air flow rate. When the stirrer speed reached maximum (1500 rpm) the DOT began to decline but this effect was arrested by control of the nutrient feed pump restricting the flow rate to prevent the DOT falling below 15% air saturation. At the end of the fermentation about 5 litres of concentrated medium had been added and the cell dry weight was greater than 100 g/litre.

The protocol provides us with a good reproducible procedure for assessment of yeast strains in the laboratory under production type conditions. The protocol is also capable of adaption to large-scale operation. We need now to consider the requirements which the proposed process places upon those designing the yeast strain. Obviously we require genetic constructs involving suitable promoter and leader sequences to cause albumin to be secreted into the medium in high yield under appropriate conditions. The scale of the proposed operation also places particular demands on the stability of the genetic construct — operation at high cell density and in large industrial fermenters implies a period of growth of 40–50 generations from seed stock to completion of the full-scale batch. Our adopted solutions to these problems are described below.

CONSTRUCTION OF A SYSTEM FOR ALBUMIN SECRETION

Secretion in yeast and other eukaryotes has been reviewed a number of times (e.g. Refs 19, 31–33). Since, however, some knowledge of this process is required for an understanding of the construction of the genetic systems which have been used to achieve albumin secretion, a brief outline of the pathway emphasising those steps which affect the genetic construct is given below.

Genes coding for secreted proteins include an N-terminal 'leader' sequence which is absent from the mature protein.[34] The first 20 or so amino acids form the 'signal' peptide[35] which serves as a recognition system by which proteins to be secreted are targetted to the membrane. In mammalian cells the process is mediated by a soluble cytoplasmic signal recognition particle[36] which binds the nascent signal sequence and arrests further translation until its subsequent transfer to the signal sequence receptor in the endoplasmic reticulum membrane via a receptor (or docking protein) in the membrane.[37] The ribosome, which is now bound to the membrane by the signal sequence of the nascent protein, proceeds to complete the synthesis of the remainder of the protein which is simultaneously secreted into the endoplasmic reticulum. It does, however, appear that in yeast post translational transport across the endoplasmic reticulum can also occur,[38] and the existence of the signal recognition particle and its receptor has yet to be established.

On completion of translation the signal peptide is cleaved by a signal peptidase[39, 40] leaving the newly synthesised protein in the endoplasmic reticulum. Folding of the protein into its native structure occurs in the endoplasmic reticulum. This process begins simultaneously with emergence of the nascent polypeptide chain[40] and is catalysed by protein disulphide isomerase and prolyl *cis-trans* isomerase.[42-44] Other processing which may occur in the ER includes core glycosylation[45, 46] and fatty acylation[47] although these modifications do not apply to albumin.

Final maturation of the protein occurs in the Golgi. Events occurring at this site include completion of glycosylation and removal of the residual leader sequences (or 'pro' peptide). It is also necessary to identify, distinguish and package separately proteins destined alternatively for containment in vacuoles, retention in the ER or Golgi, containment in secretory storage vesicles or immediate secretion. (For reviews on these processes see Refs 48–50.) Proteins destined for immediate secretion are packaged into secretory vesicles which, in yeast are believed to fuse with the growing bud tip.[19]

It is apparent from this outline of the secretion process that the main requirement of a genetic construct for the secretion of a heterologous protein is the provision of an N-terminal leader peptide including those sequences involved in signal recognition, signal peptidase action and maturation and packaging in the Golgi such that these processes will occur efficiently in yeast. One possible solution to this problem is to prefix the mature protein with a leader sequence for an authentic yeast secreted protein; invertase, acid phosphatase and α mating factor (MFα1) have all been used for this purpose. It has commonly been found that levels of heterologous proteins secreted using these leaders are higher than those observed with the natural signal sequences of the various heterologous proteins.[51]

The MFα1 pheromone leader system has been particularly intensively studied in this regard. The pheromone itself consists of a tridecapeptide which is synthesised as a precursor (pre-pro α factor) of 165 amino acids. The precursor consists of a typical N-terminal signal sequence of about 20 amino acids followed by a further 63 residues of 'pro' sequence. There follow four copies of the pheromone sequence preceded by a hexapeptide having the sequence Lys-Arg-Glu-Ala-Glu-Ala (KREAEA)[52] and separated by similar octapeptides Lys-Arg-Glu-Ala-Asp-Ala-Glu-Ala (KREADAEA). Maturation of pro α factor in the Golgi requires the action of three proteolytic enzymes: an endopeptidase coded for by the *KEX2* gene having the specificity to cleave after a pair of basic amino acids,[53] a dipeptidyl amino peptidase coded by the STE13 gene which removes the (Glu Ala)$_2$ tetrapeptide[54] and a carboxypeptidase B like activity which removed Lys-Arg pairs from the C-terminal.[55]

In some cases where the complete α factor leader has been used, removal of the (Glu Ala)$_2$ tetrapeptide has been found to be rate limiting, resulting in secretion of recombinant products with the tetrapeptide still attached or in disappointingly low yield.[56] A modified pre-pro α factor leader from which the (Glu Ala)$_2$ sequences had been deleted gave improved secretion of interferon by comparison with the full length leader.[57, 58] The author has used a similar leader, terminating in the Lys-Arg *KEX2* site and not requiring STE13 processing, to direct albumin secretion.[59]

Although originally identified in yeast from its requirement for processing type 1 yeast killer toxin and α mating factor,[54] it is now apparent that the use of a pair of basic amino acids as a processing signal for a *KEX2* like cleavage is a common motif in eukaryotic secretion. Of particular relevance to the author's work with albumin is the finding that the albumin leader is also processed by an endoprotease of the same specificity. In contrast to α factor the complete leader for albumin is much shorter being only 24 amino acids in length,[60] the first 18 of which make up the signal peptide. The pro sequence hexapeptide terminates in a *KEX2* specificity site (Arg-Arg). The involvement of a *KEX2* like processing step in albumin secretion is confirmed by the finding of circulating pro albumin in the bloodstream of individuals in which the enzyme is inhibited either by the presence of the abnormal α_1 antitrypsin Pittsburgh[61] or by an Arg $\rightarrow$ Glu mutation in the final position in the leader peptide.[62] Finally, isolated yeast *KEX2* protein has been shown to be capable of converting pro-albumin to albumin.[63]

Since similar recognition signals are used in the processing of yeast α factor and albumin leaders there seemed grounds for supposing that in this case the natural albumin leader might be correctly processed by yeast. It has been found that this is in fact so and that both leaders direct albumin secretion in yeast in similar yield. There are, however, subtle differences in the systems; when α factor leader is used to direct albumin secretion there accumulates, in addition to authentic albumin, a substantial proportion (to about 30% of the albumin level) of an incomplete albumin molecule of about 45 000 molecular weight which is lacking a C-terminal region of the molecule. The use of natural albumin leader results in the production of only one third this level of the albumin fragment.[64] The possibility that the generation of this incomplete molecule might itself be the consequence of *KEX2* acting on a potential cleavage site (Lys^{413}–Lys^{414}) within the albumin molecule was considered. However site specific mutation to abolish this site did not affect the amounts of fragment produced.[64]

In addition to the leader sequences mentioned above, three other leaders have also been examined, consisting of:

(i) The leader sequence for *Kluyervomyces lactis* killer toxin.
(ii) A hybrid leader formed by fusing the *K. lactis* killer signal sequence to the pentapeptide (SLDKR)-positions 81–85 from MFα1 leader terminating in the KEX2 site.
(iii) A hybrid formed by fusing the signal sequence of HSA to the same pentapeptide MFα1 sequence (i.e. residues 81–85).

Note that all five leaders studied terminate in a *KEX2* site. Of these the last described HSA-MFα1 fusion gave the best results in terms of albumin yield and minimal fragment production.[64]

CONSTRUCTION OF PLASMID VECTOR SYSTEMS

Having selected an appropriate leader sequence it is necessary to arrange a suitable expression system. The requirements for stability and high expression imply that a high copy number plasmid system should be advantageous.[65] Suitable promoter and transcription terminators are required respectively upstream and downstream of the coding region to control mRNA production. The promoter may be either inducible or constitutive. The major theoretical advantage of an inducible promoter is that production of any recombinant protein, even if it has no direct effect on the producer cell, must be to some extent a drain on the cell's resources and might be expected to lead to some reduction in yield or growth rate. The experience in bacterial systems is that this instability is so severe that the use of inducible promoters is obligatory. In yeast, however, the effect seems to be much less severe.

Vector systems used for cloning in yeast are most commonly based upon the $2\,\mu$m plasmid[66] which is found in nearly all strains of *S. cerevisiae*.[67, 68] They have been chosen for this purpose because of their relatively high copy number and stability.[69] In addition to the sequences for recombinant protein expression, the minimum requirements of such plasmids are: an *E. coli* origin of replication and selectable marker (commonly derived from the bacterial plasmids such as pBR322 or pUC), a yeast origin of replication and the *cis* acting STB (also known as REP3) region of the $2\,\mu$m plasmid,[70, 71] and a selectable marker which permits detection and selection for yeast transformants. Such selectable markers are commonly chosen to complement an auxotrophic require-ment in the haploid yeast host.[66] We have used pJDB207 based vectors[72] for this purpose in which the selectable marker is LEU2, such a plasmid is illustrated in Fig. 2. Plasmid bearing cells are identified as prototrophs in a leu2⁻ host background and selection for such transformants is maintained simply by growth on minimal media. Thus the media which we intended to use in the fermentation process should select for retention of the plasmid.

In agreement with other workers[69, 73] the author has found that the rate of loss of such plasmids is of the order 1–2% per generation under

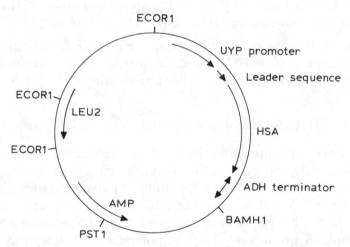

Fig. 2. Generic plasmid base for the expression/secretion of human serum albumin from haploid yeast strains. The plasmid is based on the *E. coli*/yeast shuttle vector pJDB207 and contains the LEU2 and Amp^R selection markers; yeast $2\,\mu$ and *E. coli* ori sites and an HSA expression/secretion cassette.

selective conditions. Unfortunately this is inadequate for a process such as is being considered here which will require 40–50 generations of growth from seed stock to production scale. Thus a loss of albumin yield of around 50% after 40 generations growth is observed.

It was considered that one potential cause of this instability could be that plasmids such as pJDB207 are not totally self sufficient, but require provision of the REP1 and REP2 functions in trans from an endogenous 2μm plasmid.[71, 74, 75] The homology between the recombinant and endogenous 2μm plasmid could cause recombination to occur permitting the prototrophic marker to be transferred to the endogenous plasmid. This would result in segregation of the selectable marker from the albumin (or other heterologous protein) gene. Thus the maintenance of prototrophy would no longer select for retention of the heterologous gene and the culture would become overtaken by non-albumin producing prototrophs.

A number of vectors have been described which contain the REP1 and REP2 genes and which can therefore be maintained in an otherwise cir° (2μm minus) host (for reviews see Refs 66, 68, 76). Such plasmids should in principle be more stable because in a cir° host they cannot recombine with endogenous 2μm circles. The author has sought to improve on these by designing a recombinant plasmid which resembles the 2μm as closely as possible apart from the presence of the heterologous gene, the selectable marker and their respective transcription control sequences.[77] These were positioned as closely together as possible between the STB region and the origin of replication. In addition, the plasmid has been designed in such a way that the bacterial sequences, which are of course essential for its initial construction in *E. coli*, are excised from the plasmid after transfer to yeast.

The design of such a plasmid is illustrated in Fig. 3. The 2μm plasmid is commonly represented in a 'figure of eight' form as in Fig. 3 to emphasise the presence of two antiparallel homologous sequences of 599bp containing an XbaI site.[78, 79] Recombination between these two forms is catalysed by the FLP gene product,[80, 81] which results in the plasmid existing in two alternate equimolar forms in which the two unique regions separated by the inverted repeats are in opposite orientation to each other.[78, 82] This recombination process is thought to have an important role in plasmid copy number amplification because recombination occurring shortly after replication initiation can result in the production of many copies of the plasmid from a single initiation event.[83, 84] Repression of FLP by the combined action of the REP1 and REP2 gene products[85-87] is thought to regulate this process resulting in stable copy number of about 50 per haploid genome. The author

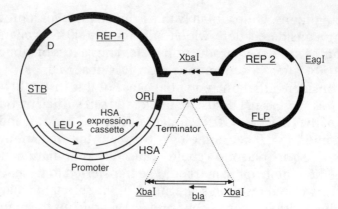

Fig. 3. Modified $2\,\mu$m plasmid for stable expression of albumin in yeast. Bacterial sequences including *E. coli* origin of replication and β-lactamase marker (bla) are flanked by two 74bp repeat sequences containing the FLP recognition target.

has introduced a 74bp fragment of part of the repeated sequence, containing the FLP recognition target or FRT[88] attached to an *E. coli* replication origin and ampicillin resistance marker. This sequence was then introduced at the XbaI site (which terminates the 74bp sequence) in the 599bp repeat, so that the bacterial sequences are now positioned between a pair of parallel 74bp repeats containing the FRT sequence. This plasmid is then capable of replication and manipulation in *E. coli*. The HSA expression cassette was introduced into the modified $2\,\mu$m plasmid between the replication origin and STB region with the LEU2 gene positioned as close as possible to the HSA sequences so as to minimise the risk of their subsequent segregation.

On transfer of the recombinant plasmid into a cir° yeast FLP catalysed recombination can occur between the two parallel homologous FRT regions leading to the excision of the bacterial sequences and the restoration of the native 599bp repeat. The bacterial sequences are no longer capable of replication in the yeast cell and are lost leaving the complete $2\,\mu$m plasmid containing only the minimum additions necessary for its selection and albumin expression.

These plasmid systems have been found to be remarkably stable, even when containing HSA expression cassettes with constitutively acting promoters. The author has been able to maintain some haploid yeast strains (leu2⁻ cir° before transformation) transformed with such plasmids for hundreds of generations in continuous culture on carbon limiting minimal media without observing segregation of the LEU2 marker from the albumin secretion phenotype. With appropriate promoters it has been possible to obtain sufficiently high yields for an economically viable process and these high yields have been maintained in intensive fed batch fermentation to high dry weight, as discussed above. The author believes that the combination of high albumin yield, intensive fermentation and high stability provide us with a basis from which to develop an albumin production process.

ACKNOWLEDGEMENT

This chapter includes results from many of the author's colleagues at Delta. He wishes to thank particularly Peter Clarke and David Mead for the fermentation work, Simon Chinery and Ed Hinchliffe who developed the plasmid vector system, and Andrew Goodey, Darrell Sleep and Graham Belfield for the studies on the various leader systems.

REFERENCES

1. Peters, T. Jr, Serum Albumin. *Adv. Prot. Chem.,* **37** (1985) 161–245.
2. Kragh-Hansen, U., Molecular aspects of ligand binding to serum albumin. *Pharmacol Rev.,* **33** (1981) 17–57.
3. Klausner, A., Adjustment in the blood fraction market. *Bio/Technology,* **3** (1985) 119–25.
4. Jeans, E. R. A., Marshall, P. J. & Lowe, C. R., Plasma protein fractionation. *Trends Biotechnol.,* **3** (1985) 267–70.
5. Meloun, B., Moravek, L. & Kostka, V., Complete amino acid sequence of human serum albumin. *FEBS Lett.,* **58** (1975) 134–7.
6. Glaumann, H., Albumin secretory pathway in the hepatocyte. In *Albumin Structure, Biosynthesis and Function,* ed. T. Peters & I. Sjoholm. Pergamon Press, Oxford, 1978, pp. 41–50.
7. Andersson, L. O., Reduction and re-oxidation of the disulfide bonds of bovine serum albumin. *Arch. Biochem. Biophys.,* **133** (1969) 277–85.
8. Kenten, J., Helm, B., Ishizaka, T., Cattini, P. & Gould, H., Properties of a human immunoglobulin epsilon-chain fragment synthesized in *Escherichia coli. Proc. Natl. Acad. Sci. USA,* **81** (1984) 2955–9.
9. Emtage, J. S., Angal, S., Doel, M. T., Harris, T. J. R., Jenkins, B., Lilley, G. & Lowe, P. A., Synthesis of calf prochymosin (prorennin) in *Escherichia coli. Proc. Nat. Acad. Sci., USA,* **80** (1983) 3671–5.
10. Simons, G., Remaut, E., Allet, B., Devos, R. & Fiers, W., High-level expression of human interferon gamma in *Escherichia coli* under control of the pL promoter of bacteriophage lambda. *Gene,* **28** (1984) 55–64.
11. Latta, M., Knapp, M., Sarmientos, P., Brefort, G., Becquart, J., Guerrier, L., Jung, G. & Mayaux, J. F., Synthesis and purification of mature human serum albumin from *Escherichia Coli. Bio/Technology,* **5** (1987) 1309–14.
12. Hinchliffe, E., Kenny, E. & Leaker, A. J., Novel products from surplus yeast via recombinant DNA technology. In *European Brewery Convention*

Symposium on Brewers Yeast Monograph XII. Verlag Hans Carl (Brauwelt-Verlag), Nurnberg, 1987, pp. 139–54.

13. Burton, S. J., Quirk, A. V. & Wood, P. C., Refolding human serum albumin at relatively high protein concentration. *Eur. J. Biochem.,* **179** (1989) 379–87.

14. Gray, G. L., Baldridge, J. S., McKeown, K. S., Heyneker, H. L. & Chang, C. N., Periplasmic production of correctly processed human growth hormone in *Escherichia coli*: natural and bacterial signal sequences are interchangeable. *Gene,* **39** (1985) 247–54.

15. Becker, G. W. & Hoiung, H. M., Expression, secretion and folding of human growth hormone in *Escherichia coli*. Purification and characterization. *FEBS Lett.,* **204** (1986) 145–50.

16. Murray, R. G. E., Steed, P. & Elson, H. E., The location of the mucopeptide in sections of the cell wall of *Escherichia coli* and other Gram ⁻ve bacteria. *Can. J. Microbiol.,* **11** (1965) 547–60.

17. Novotny, A., Molecular aspects of endotoxic reactions. *Bacteriol. Rev.,* **33** (1969) 72–98.

18. Saunders, C. W., Schmidt, B. J., Mallonee, K. L. & Guyer, M. S., Secretion of human serum albumin from *Bacillus subtilis. J. Bacteriol.,* **169** (1987) 2917–25.

19. Schekman, R. & Novick, P., The secretory process and yeast cell-surface assembly. In *The Molecular Biology of the Yeast Saccharomyces: Metabolism & Gene Expression*, ed. J. N. Strathern, E. W. Jones & J. R. B. Broach. Cold Spring Harbor Laboratory, Cold Spring Harbor, NY, 1982, pp. 361–98.

20. Burrows, S., Baker's yeast. In *The Yeasts* Vol. 3, ed. A. H. Rose & J. S. Harrison. Academic Press, New York, 1970, pp. 349–420.

21. Fiechter, A., Fuhrmann, G. F. & Kappeli, O., Regulation of glucose metabolism in growing yeast cells. *Adv. Microb. Physiol.,* **22** (1981) 123–83.

22. Kingsman, S., Wilson, M. J., Cousens, D. J. & Hinchliffe, E., Yeast promoter. British Patent Application GB 8720396, 1986.

23. Condra, J. H., Ellis, R. W., Jones, R. E. & Schultz, L. D., Expression of recombinant proteins in yeast. European Patent Application EP0234862, 1987.

24. De Deken, R. H., The Crabtree effect: a regulatory system in yeast. *J. Gen. Microbiol.,* **44** (1966) 149–56.

25. Sonnlietner, B. & Kaeppeli, O., Growth of *Saccharomyces cerevisiae* is controlled by its limited respiratory capacity: formulation and verification of a hypothesis. *Biotechnol. Bioeng.,* **28** (1986) 927–37.

26. Crabtree, H. G., Observations on the carbohydrate metabolism of tumors. *Biochem. J.,* **23** (1929) 536–45.

27. von Meyenburg, K. H., Energetics of the budding cycle of *Saccharomyces cerevisiae* during glucose limited aerobic growth. *Arch. Microbiol.,* **66** (1969) 289–303.

28. Fiechter, A. & von Meyenburg, K. H., Automatic analysis of gas exchange in microbial systems. *Biotech. Bioengng,* **10** (1968) 535–49.

29. Lloyd, D., Kristensen, B. & Degn, H., Glycolyis and respiration in yeasts: the effect of ammonium ions studied by mass spectrometry. *J. Gen. Microbiol.,* **129** (1983) 2125–7.

30. Wang, H. Y., Cooney, C. L. & Wang, D. I. C., Computer control of bakers' yeast production. *Biotechnol. Bioengng,* **21** (1979) 975–95.

31. Walter, P., Signal recognition. Two receptors act sequentially. *Nature,* **328** (1987) 763–4.
32. Verner, K. & Schatz, G., Protein translocation across membranes. *Science,* **241** (1988) 1307–13.
33. Walter, P. & Lingpappa, V., Mechanism of protein translocation across the endoplasmic reticulum membrane. *Ann. Rev. Cell Biol.,* **2** (1986) 499–516.
34. Rapoport, T. A., Protein translocation across and integration into membranes. *CRC Crit. Rev. Biochem.,* **20** (1986) 73–137.
35. Blobel, G. & Dobberstein, B., Transfer of proteins across membranes. I. Presence of proteolytically processed and unprocessed nascent immunoglobulin light chains on membrane-bound ribosomes of murine myeloma. *J. Cell Biol.,* **67** (1975) 835–51.
36. Walter, P., Ibrahimi, I. & Blobel, G., Translocation of proteins across the endoplasmic reticulum. I. Signal recognition protein (SRP) binds to in-vitro-assembled polysomes synthesizing secretory protein. *J. Cell Biol.,* **91** (1981) 545–50.
37. Gilmore, R., Blobel, G. & Walter, P., Protein translocation across the endoplasmic reticulum. I. Detection in the microsomal membrane of a receptor for the signal recognition particle. *J. Cell Biol.,* **95** (1982) 463–9.
38. Schatz, G., Protein translocation: a common mechanism for different membrane systems. *Nature,* **321** (1986) 108–9.
39. Baker, R. K., Bentivoglio, G. P. & Lively, M. O., Partial purification of microsomal signal peptidase from hen oviduct. *J. Cell Biochem.,* **32** (1986) 193–200.
40. Bohni, P. C., Deshaies, R. J. & Schekman, R. W., SEC11 is required for signal peptide processing and yeast cell growth. *J. Cell Biol.,* **106** (1988) 1035–42.
41. Peters, T. & Davidson, L. K., The biosynthesis of rat serum albumin. In vivo studies on the formation of the disulfide bonds. *J. Biol. Chem.,* **257** (1982) 8847–53.
42. Lang, K. & Schmid, F. X., Protein-disulphide isomerase and prolyl isomerase act differently and independently as catalysts of protein folding. *Nature,* **331** (1988) 453–5.
43. Fuchs, S., De Lorenzo, F. & Anfinsen, C. B., Studies on the mechanism of the enzymic catalysis of disulfide interchange in proteins. *J. Biol. Chem.,* **242** (1967) 398–402.
44. Fischer, G., Bang, H. & Mech, C., Determination of enzymatic catalysis for the *cis-trans*-isomerization of peptide binding in proline-containing peptides. *Biochim. Biophys. Acta,* **43** (1984) 1101–11.
45. Julius, D., Schekman, R. & Thorner, J., Glycosylation and processing of prepro-alpha-factor through the yeast secretory pathway. *Cell,* **36** (1984) 309–18.
46. Rothman, J. E. & Lodish, H. F., Synchronised transmembrane insertion of glycosylation of a nascent membrane protein. *Nature,* **269** (1977) 775–80.
47. Berger, M. & Schmidt, F., Protein fatty acyltransferase is located in the rough endoplasmic reticulum. *FEBS Lett.,* **187** (1985) 289–94.
48. Pfeffer, S. R. & Rothman, J. E., Biosynthetic protein transport and sorting by the endoplasmic reticulum and Golgi. *Ann. Rev. Biochem.,* **56** (1987) 829–52.
49. Rose, J. K. & Doms, R. W., Regulation of protein export from the endoplasmic reticulum. *Ann. Rev. Cell Biol.,* **4** (1988) 257–88.

50. Schekman, R., Protein localization and membrane traffic in yeast. *Ann. Rev. Cell Biol.,* **1** (1985) 115–43.
51. Das, R. C. & Shultz, J. L., Secretion of heterologous proteins from *Saccharomyces cerevisiae. Biotechnol. Progress,* **3** (1987) 43–8.
52. Kurjan, J. & Herskowitz, I., Structure of a yeast pheromone (MF alpha): a putative alpha-factor precursor contains four tandem copies of mature alpha-factor. *Cell,* **30** (1982) 933–43.
53. Julius, D., Brake, A., Blair, L., Kunisawa, R. & Thorner, J., Isolation of the putative structural gene for the lysine–arginine-cleaving endopeptidase required for processing of yeast prepro-alpha-factor. *Cell,* **37** (1984) 1075–89.
54. Julius, D., Blair, L., Brake, A., Sprague, G. & Thorner, J., Yeast alpha factor is processed from a larger precursor polypeptide: the essential role of a membrane-bound dipeptidyl aminopeptidase. *Cell,* **32** (1983) 839–52.
55. Achstetter, T. & Wolf, D. H., Hormone processing and membrane-bound proteinases in yeast. *EMBO J.,* **4** (1985) 173–7.
56. Bitter, G. A., Chen, K. K., Banks, A. R. & Lai, P.-H., Secretion of foreign proteins from *Saccharomyces cerevisiae* directed by alpha-factor gene fusions. *Proc. Nat. Acad. Sci., USA,* **81** (1984) 5330–4.
57. Zsebo, K. M., Lu, H.-S., Fieschko, J. C., Goldstein, L., Davis, J., Duker, K., Suggs, S. V., Lai, P.-H. & Bitter, G. A., Protein secretion from *Saccharomyces cerevisiae* directed by alpha-factor leader region. *J. Biol. Chem.,* **261** (1986) 5858–65.
58. Carter, B. L. A., Doel, S., Goodey, A. R., Piggot, J. R. & Watson, M. E. W., Secretion of mammalian polypeptides by yeast. *Microbiol. Sci.,* **3** (1986) 23–7.
59. Sleep, D., Belfield, G. P. & Goodey, A. R., The secretion of human serum albumin by *Saccharomyces cerevisiae. Yeast,* **4** (1988 Spec. Iss.) S168.
60. Dugaiczyk, A., Law, S. W. & Dennison, O. E., Nucleotide sequence and the encoded amino acids of human serum albumin mRNA. *Proc. Nat. Acad. Sci., USA,* **79** (1982) 71–5.
61. Brennan, S. O., Owen, M. C., Boswell, D. R., Lewis, J. H. & Carrell, R. W., Circulating proalbumin associated with a variant proteinase inhibitor. *Biochim. Biophys. Acta,* **802** (1984) 24–8.
62. Brennan, S. O. & Carrell, R. W., A circulating variant of human proalbumin. *Nature,* **274** (1978) 908–9.
63. Bathurst, I. C., Brennan, S. O., Carrell, R. W., Cousens, L. S., Brake, A. I. & Barr, P. J., Yeast KEX2 protease has the properties of a human proalbumin converting enzyme. *Science,* **235** (1987) 348–50.
64. Sleep, D., Belfield, G. P. & Goodey, A. R., The secretion of human serum albumin by the yeast *Saccharomyces cerevisiae. Bio/Technology,* (submitted).
65. Futcher, A. B. & Cox, B. S., Maintenance of the 2-micron circle plasmid in populations of *Saccharomyces cerevisiae. J. Bacteriol.,* **154** (1984) 612–22.
66. Broach, J. R., Construction of high copy yeast vectors fusing 2-micron circle sequences. *Methods Enzymol.,* **101** (1983) 307–25.
67. Tubb, R. S., 2-micron DNA plasmid in brewery yeasts. *J. Inst. Brew.,* **86** (1980) 78–80.
68. Broach, J. R., The yeast plasmid 2-micron circle. In *The Molecular Biology of the Yeast Saccharomyces Life Cycle & Inheritance,* ed. J. N. Strathern, E. W. Jones & J. R. B. Broach. Cold Spring Harbor Laboratory, Cold Spring Harbor, NY, 1981, pp. 445–70.

69. Futcher, A. B. & Cox, B. S., Copy number and the stability of 2-micron circle-based artificial plasmids of *Saccharomyces cerevisiae. J. Bacteriol.*, **157** (1984) 283–90.
70. Jayaram, M., Sutton, A., & Broach, J. R., Properties of REP3: a *cis*-acting locus required for stable propagation of the *Saccharomyces cerevisiae* plasmid 2-microns circle. *Mol. Cell Biol.*, **5** (1985) 2466–75.
71. Kikuchi, Y., Yeast plasmid requires a *cis*-acting locus and two plasmid proteins for its stable maintenance. *Cell*, **35** (1983) 487–93.
72. Beggs, J. D., Multiple-copy yeast plasmid vectors. *Alfred Benzon Symposium*, **16** (1981) 383–95.
73. Murray, A. W. & Szostak, J. W., Pedigree analysis of plasmid segregation in yeast. *Cell*, **34** (1983) 961–70.
74. Cashmore, A. M., Albury, M. S., Hadfield, C. & Meacock, P. A., Genetic analysis of partitioning functions encoded by the 2 μm circle of *Saccharomyces cerevisiae. Mol. Gen. Genet.*, **203** (1986) 154–62.
75. Jayarum, M., Li, Y. Y. & Broach, J. R., The yeast plasmid 2 μm circle encodes components required for its high copy propagation. *Cell*, **34** (1983) 95–104.
76. Futcher, A. B., The 2-micron circle plasmid of *Saccharomyces cerevisiae. Yeast*, **4** (1988) 27–40.
77. Chinery, S. A. & Hinchliffe, E., The stable maintenance of 2-micron plasmids in yeast: a new class of vector. *Yeast*, **4** (1988 Spec. Iss.) S123.
78. Guerineau, M., Grandchamp, C. & Slonimski, P., Circular DNA of a yeast episome with two inverted repeats: structural analysis by a restriction enzyme and electron microscopy. *Proc. Nat. Acad. Sci., USA*, **73** (1976) 3030–4.
79. Hartley, J. L. & Donelson, J. E., Nucleotide sequence of the yeast plasmid. *Nature*, **286** (1980) 860–5.
80. Blanc, H., Gerbaud, C., Slonimski, P. & Guerineau, M., Stable yeast transformation with chimeric plasmids using a 2-micron-circular DNA-less strain as a recipient. *Mol. Gen. Genet.*, **176** (1979) 335–42.
81. Broach, J. R. & Hicks, J. B., Replication and recombination functions associated with the yeast plasmid, 2 μ circle. *Cell*, **21** (1980) 501–8.
82. Gubbins, E. J., Newlon, C. S., Kann, M. D. & Donelson, J. E., Sequence organization and expression of a yeast plasmid DNA. *Gene*, **1** (1977) 185–207.
83. Fuchter, A. B., Copy number amplification of the 2-micron circle plasmid of *Saccharomyces cerevisiae. J. Theor. Biol.*, **119** (1986) 197–204.
84. Volkert, F. C. & Broach, J. R., Site-specific recombination promotes plasmid amplification in yeast, *Cell*, **46** (1986) 541–50.
85. Murray, J. A., Scarpa, M., Rossi, N. & Cesareni, G., Antagonistic controls regulate copy number of the yeast 2 μ plasmid. *EMBO J.*, **6** (1987) 4205–12.
86. Reynolds, A. E., Murray, A. W. & Szostak, J. W., Roles of the 2-micron gene products in stable maintenance of the 2-micron plasmid of *Saccharomyces cerevisiae. Mol. Cell Biol.*, **7** (1987) 3566–73.
87. Som, T., Armstrong, K. A., Volkert, F. C. & Broach, J. R., Autoregulation of 2-micron circle gene expression provides a model for maintenance of stable plasmid copy levels. *Cell*, **52** (1988) 27–37.
88. McLeod, M., Craft, S. & Broach, J. R., Identification of the crossover site during FLP-mediated recombination in the *Saccharomyces cerevisiae* plasmid 2-microns circle. *Mol. Cell Biol.*, **6** (1986) 3357–67.

Chapter 6

STRATEGIES FOR EXPRESSING CLONED GENES IN MAMMALIAN CELLS

MARY M. BENDIG

Medical Research Council, Collaborative Centre, Mill Hill, London, UK

INTRODUCTION

The ability to express cloned genes in mammalian cell lines has proved to be a powerful technology. It is now possible to produce a rare protein of some scientific or therapeutic interest in sufficient quantity to be able to study it and/or produce it commercially. With the genes coding for interesting proteins cloned and expressable, the future holds the possibility for genetically modifying natural proteins to design and produce new proteins as required. The best example, to date, of a cloned gene expressed in mammalian cells to produce a human therapeutic agent is tissue plasminogen activator (tPA). Interestingly, there are already underway projects to engineer natural tPA protein to create new 'second-generation' tPA proteins with altered properties such as longer half-life and increased substrate specificity[1] (see Chapter 8, this volume).

In the early days of Biotechnology (less than 10 years ago), much emphasis was put on using *E. coli*, or other simple cell types, to produce proteins in bulk from cloned human genes.[2] Many problems, however, were encountered in producing functional eukaryotic proteins in bacterial cells. Most importantly, bacterial cells do not carry out post-translational modifications such as glycosylation, amidation, phosphorylation, and cleavage of protein precursors. Producing eukaryotic proteins in mammalian cells, where correct post-translational processing generally occurs, clearly solves many of these problems. The realization that, in many cases, bacterial expression systems would be unsatisfactory has spurred the development of better mammalian cell expression systems. Coupled with the development of better mammalian expression systems has been the development of more efficient means for growing mammalian cells for large-scale production (see Chapter 13, this volume).

BASIC STRATEGIES

The strategies for expressing cloned genes in mammalian cells have exploited a few basic principles to achieve high levels of protein production. For example, more foreign gene copies per cell will, in general, lead to higher levels of protein production. Two distinct methods of achieving high gene copy number, and therefore high production levels, have been employed. In the first method, the number of copies of the protein-coding gene is increased by co-amplification of the number of desired protein-coding genes and of a selectable, amplifiable gene (see review by Bebbington and Hentschel[3]). The most popular mammalian expression system based on co-amplification uses vectors containing the dihydrofolate reductase (DHFR) gene as the selectable gene and Chinese hamster ovary (CHO) cells as the host cell line. Cells can be selected that contain up to 1000 copies per cell of the vector DNA.[4]

The second method of achieving elevated gene copy number, and increased protein production, relies on viral DNA replication to achieve high copy numbers. A widely used example of this approach is the expression system based on bovine papillomavirus (BPV) vectors in rodent fibroblasts such as mouse C127 cells. When introduced into C127 cells, BPV-based DNA vectors replicate and are often maintained in the cells either episomally or tandemly integrated into mouse chromosomal DNA. The result is BPV-transformed C127 cells carrying up to 200 foreign protein-coding genes per cell (see review by Stephens and Hentschel[5]).

In addition to increasing gene copy number, the level of foreign protein production can be increased by linking the protein-coding sequences to a strong promoter/enhancer sequence that will give high levels of transcription and, therefore, high levels of protein production. Many mammalian expression vectors use well-characterized viral promoters/enhancers, such as those from SV40,[6] for this purpose.

The DNA vector and its design are only one half of the equation for a good mammalian cell expression system. The cell line used to express the foreign gene and to produce the desired protein product provide the other half. The ideal cell line should be easy to transfect, should stably maintain the foreign DNA with an unrearranged sequence, and should have the necessary cellular components for efficient transcription, translation, post-translational modification, and secretion. Unfortunately, our ability to manipulate the characteristics of potential cell lines to create the ideal cell line for expression purposes is much more limited than our ability to manipulate DNA expression vectors. The mammalian cell lines routinely used for expression studies are almost always cell lines that have been widely used for experimental purposes. They grow easily in culture, and experience has demonstrated reasonably efficient means of introducing DNA vectors into the cells. Experience has also shown that the cells are usually capable of producing functional protein products from the introduced foreign genes. Little has been done to optimize or improve the recipient or host cell lines themselves. Recently some effort has been made to choose host cell lines that might be predicted to possess the traits necessary to be good producers of the desired protein product. For example, myeloma cell lines are considered to be the cell lines of choice for expressing genetically-manipulated immunoglobulin genes.[7] Although there may be significant advantages in expressing antibody genes in myeloma cells, functionally active antibody can be produced and secreted from both lymphoid and non-lymphoid cells (see Tables 1 and 2).

THREE EXPRESSION SYSTEMS

At present, two mammalian cell expression systems have been most successful and will be the systems primarily used to make recombinant DNA products introduced during 1988–1990.[8] These systems are DHFR gene amplification vectors in CHO cells and BPV vectors in mouse C127 cells. As already discussed, both systems rely on high gene copy number to achieve high levels of protein production. Another mammalian cell expression system thought to have great potential is based on using murine myeloma cell lines (see Fig. 1 and Table 3).

I. Gene Amplification Vectors in CHO cells

The most widely used, and arguably most successful mammalian expression system with respect to levels of protein production achieved, employs a DHFR minus CHO cell line and a DNA vector containing the DHFR gene and some strong promoter to drive the gene to be expressed for protein production. The gene to be expressed is cloned into the vector, and CHO cells are transfected with the resultant DNA. Cells taking up the DNA, and therefore expressing the DHFR gene, are selected as foci able to grow in the cytotoxic drug, methotrexate (MTX). Primary resistant cells are then grown in progressively higher concentrations of MTX. This protocol selects for cells with amplified copies of the DHFR gene. Since the gene to be expressed is adjacent to the DHFR gene, and since the amplification unit is large, amplification of the DHFR gene

Table 1
Examples of Expressing Engineered Antibody Genes in Myeloma Cell Lines

Cell	Vector	Antibody genes	Production levels	References
Non-producing mouse P3 line	Genomic mouse Ig promoters, *gpt* or *neo* for selection	Genomic mouse variable (anti-phosphocholine) plus genomic human constant regions	Equal to low level mouse hybridoma line	40
SP2/0–Ag14	Genomic mouse Ig promoters, *gpt* or *neo* for selection	Genomic mouse variable (anti-TNP) plus genomic human constant regions	5 μg/ml	41, 42
Sp2/0–Ag14, P3X63Ag8.653	Genomic mouse Ig promoters, *gpt* or *neo* for selection	Genomic mouse variable (anti-human breast carcinoma, B6.2) plus genomic human constant regions	1 μg/ml/10^6 cells/24 h	43
Sp2/0–Ag14	Genomic mouse Ig promoters, *gpt* or *neo* for selection	Genomic mouse variable (anti-human colorectal carcinoma, C017-1A) plus genomic human constant regions	15 μg/ml	44, 45
P3X63Ag8.653	Genomic mouse Ig promoters plus cloned human IgH enhancer, *gpt* or *neo* for selection	Genomic mouse variable (anti-cALLA) plus genomic human constant regions	30 μg/ml	46
Sp2/0–Ag14	SV40 early promoter plus cloned mouse IgH enhancer, *neo* for selection	Mouse variable region cDNAs (anti-human carcinoma, L6) plus human constant region cDNAs	1 μg/ml	47, 48
Rat YO	Cloned mouse IgH promoter plus cloned mouse IgH enhancer, *gpt* or *neo* for selection	Synthetic human variable regions (anti-CAMPATH-1) plus genomic human constant regions	10 μg/ml	49

Table 2
Examples of Expressing Engineered Antibody Genes in Non-lymphoid Cell Lines

Cell	Vector	Antibody genes	Production levels	References
CV-1 (African green monkey), Hela	SV40 promoter (no expression with mouse Ig promoter)	Genomic mouse kappa light chain	Efficient transient expression	50
CHO *dhfr*⁻	SV40 early promoter, *dhfr* for selection	Mouse light and heavy cDNAs (anti-creatine kinase)	35 ng/ml/10^6 cells/24 h (no gene amplification)	51
CHO	HCMV promoter and enhancer, *gpt* or *neo* for selection	Mouse variable region cDNAs (anti-human breast carcinoma, B72.3) plus genomic human constant regions	100 μg/ml	52

Mary M. Bendig

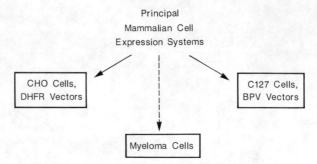

Fig. 1. Two proven mammalian cell expression systems and a possible new expression system for the future.

generally results in co-amplification of the gene to be expressed (see Refs 9 and 10).

Gene amplification is a general process and there are at least 12 examples of genes that are selected in an amplified state when the cells are exposed to certain toxic drugs.[11] Other genes besides DHFR that have been used as both a selectable marker and an amplifiable gene in CHO cells are the adenosine deaminase (ADA) gene and glutamine synthetase (GS) gene. The drugs used for ADA selection and amplification are 2′-deoxycoformycin plus cyctotoxic concentrations of adenosine or an analogue, 9′-β-D-xylofuranosyl adenine.[12] For GS selection and amplification, the cells, which must be able to take up glutamic acid from the medium, are treated with the inhibitor methionine sulphoximine (MSX).[3] ADA and GS expression vectors may offer some advantages over DHFR vectors in that ADA⁻ or GS⁻ host cells are not necessary and so these expression vectors can be introduced and amplified in a wider variety of mammalian cells. In some cases, however, DHFR vectors have also been used in wild-type host cell lines.[13-15]

A few examples of proteins that have been produced in high levels by CHO cells with amplifiable DHFR expression vectors are hepatitis B

Table 3
Performance of Principal Mammalian Cell Expression Systems

I.   CHO cells, DHFR gene amplification vector
      Tissue plasminogen activator (tPA) 66 $\mu g/10^6$ cells/day

II.   C127 cells, BPV vector
      Human growth hormone (hGH) 18 $\mu g/10^6$ cells/day

III. Murine myeloma cells
      Engineered antibody molecules 10 to 30 $\mu g/ml$ ($\sim 10^6$ cells/ml)
      Potential for high productivity: murine monoclonal antibody 8 to 130 $\mu g/10^6$ cells/day

---

surface antigen (HBsAg),[16] human tissue plasminogen activator (tPA),[17] and human interferon-γ (IFN-γ).[18] Table 4 summarizes the production levels achieved after selection for amplification. Interestingly, the three different proteins were produced at roughly the same levels, particularly with respect to the number of molecules produced per cell. The promoter/enhancer combination used to drive the protein-coding sequences differed in each example, but in all cases consisted of a viral promoter and enhancer known to give high levels of transcription. In the three examples cited, the proteins, as produced in CHO cells, were very similar to the native protein. The CHO-produced human tPA was secreted, had a specific activity approximately the same as native tPA, and was glycosylated in a similar but not identical manner.[17] The CHO-produced human IFN-γ, like IFN-γ from white blood cells, was a mixture of two glycosylated species with traces of an unglycosylated molecule. Also, unlike *E. coli*-produced IFN-γ, the CHO-produced IFN-γ was processed to remove all the cysteines and thus is identical to the native protein in this respect.[18] The HBsAg particles secreted from CHO cells were identical in size and density to the 22 nm particles from human plasma. Two of the three polypeptides in the CHO-produced HBsAg particles were antigenically indistinguishable from human plasma HBsAg particles and immunogenic in chimpanzees.[16, 19] In conclusion, CHO cells with amplified gene copy number can produce high levels, up to 5×10^{14} molecules/10^6 cells/day, of correctly, or very nearly correctly, processed human protein.

The major advantage of CHO/gene amplification expression systems is that amplifying the gene copy number results in a roughly proportional increase in protein production. The major disadvantage of gene amplification expression systems is the questionable stability of cell lines with artificially-induced high gene copy numbers. Amplification of genes in CHO cells tend to be those that are already integrated into specific, normal chromosomes rather than present on small double minute chromosomes (DMs). Amplification of genes integrated into large chromosomal DNA is more stable than amplification of DMs (see review by Stark and Wahl[10]). Selecting CHO cells in high levels of MTX to obtain cell lines that contain as many as 1000 copies of the DHFR gene per cell appears to give rise to a variety of cell lines some of which are relatively stable in the absence of selective pressure and others which are not. Some of the tPA-producing CHO cell lines were stable in the absence of MTX and continued to produce high levels of tPA for over 30 cell doublings. Other clones fell sharply in tPA production when propagated in the absence of MTX.[17] With the HBsAg-producing CHO cells, cell lines resistant to the highest levels of MTX, and therefore the

Mary M. Bendig

Table 4

Examples of Protein Production Using Amplifiable DHFR Expression Vectors in CHO Cells

| Protein produced | Expression vector | | Gene copies/cell | Production levels[a] | | References |
	Promoter	Enhancer		µg/10⁶ cells/day	Molecules/10⁶ cells/day	
Hepatitis B surface antigen	HBsAg	SV40	400	15	3.3×10^{14}	16
Human tissue plasminogen activator	Adenovirus major late	SV40	100-fold increase after amplification	66	5.8×10^{14}	17
Human interferon-γ	SV40 early	duplicated SV40	20–50-fold increase after amplification	10–20	2.8×10^{14}	18

[a]The following assumptions were made. The average molecular weight of hepatitis B surface antigen is 27 000 D.[16] For tissue plasminogen activator, the specific activity is 90 000 units/mg and the molecular weight is 68 000 D.[53] The specific activity of human interferon-γ is 1 to 2×10^{8} units/mg.[18]

highest producers of HBsAg, were unstable in the absence of MTX with production levels declining. Clones resistant to less MTX and producing intermediate levels of HBsAg were stable in the absence of MTX.[16] Any cell line destined to be used for large-scale commercial production of some foreign protein should be a stable producer, or at least stable over a production run. Selective media are generally not suitable for use in large-scale production, either because of cost or toxicity, and so the protein-producing cell lines must be stable producers in the absence of the selective pressure to retain high gene copy numbers.

II. Bovine Papilloma Virus Vectors

BPV DNA will transform rodent fibroblasts in culture. The viral DNA does not usually integrate into the cellular genome but instead is maintained episomally as circular DNA with up to several hundred copies per cell. The high copy number and episomal state of BPV DNA in transformed cells have been exploited to create BPV-based mammalian cell expression vectors (see review by DiMaio[20]). The most successful examples of foreign protein production in BPV-transformed cells employ C127 cells, a fibroblastic mouse mammary tumour cell line,[21] as the host cell. A BPV expression vector generally contains the BPV DNA necessary for cell transformation and high copy number, the foreign gene to be expressed together with the necessary transcriptional and translational regulatory sequences. In addition there is some bacterial DNA for replication and selection in bacteria. Some BPV-based expression vectors also carry a gene, such as metallothionein, to allow for easy selection of successfully transfected host cells and/or to extend the host range. Although BPV DNA itself exists as a stable episome in transformed C127 cells,[22] the complex BPV-based expression vectors often become integrated into the host cell genome (see Ref. 5). For the purpose of efficient production of foreign proteins, integration is not a problem as long as there is no significant gene rearrangement and the high copy number is maintained.

BPV-based vectors have been used to produce at least 26 different foreign proteins.[5] High levels of protein production are believed to be due partly to high gene copy number. In addition, BPV vectors contain the BPV enhancer elements which contribute to high levels of expression.[23] Table 5 lists four examples of proteins produced at relatively high levels in BPV-transformed mouse C127 cells. In all four examples, the BPV vectors are very similar with the mouse metallothionein-I promoter used to express the protein-coding cDNA. Other promoters, such as the SV40

Mary M. Bendig

Table 5
Examples of Protein Production Using BPV Vectors in C127 Cells

Protein product	Expression vector		Gene copies/cell	Production levels[a]		References
	Promoter	Enhancer		μg/10^6 cells/day	Molecules/10^6 cells/day	
Human growth hormone	Mouse MMT-I	Mouse MMT-I	10–100	18	6×10^{14}	25
Hepatitis B surface antigen	Mouse MMT-I	Mouse MMT-I	5–180	1	$2 \cdot 2 \times 10^{13}$	26
Human chorionic gonadotropin α-subunit	Mouse MMT-I	Mouse MMT-I	ND[b]	0·3	$8 \cdot 2 \times 10^{12}$	54
Human tissue Plasminogen activator	Mouse MMT-I	Mouse MMT-I	10–150	3·1	$2 \cdot 7 \times 10^{13}$	27

[a]In addition to the assumptions made in Table 1, the molecular weight of human chorionic gonadotropin was taken as 22 000 D.[54]
[b]Not determined.

early promoter, have been used but less successfully.[24] The highest reported production levels from the BPV vector/C127 cell expression system are for human growth hormone (hGH).[25] On a molecules/cell basis, the BPV-transformed C127 cells produce as much hGH as the best reported examples of foreign protein production in CHO/DHFR expression systems (see Tables 4 and 5). The HBsAg has been produced by both the CHO/DHFR and BPV/C127 expression systems (see Tables 4 and 5). The production levels are about 15-fold higher in the gene-amplified CHO cells. The higher production levels in CHO cells are probably due to the higher gene copy number achieved with this system.

The proteins produced in C127 cells, like those produced in CHO cells, are generally correctly processed and biologically active. The HBsAg produced in C127 cells is secreted as a 22 nm particle which corresponds physically, biochemically, and immunologically to the HBsAg particles isolated from human serum.[26] The hGH protein is processed and secreted by the C127 cells to give a protein migrating with authentic hGH on SDS-polyacrylamide gels.[25] The tPA protein secreted by C127 cells is correctly processed, has been glycosylated, and is biologically active as measured in fibrinolytic assays.[27, 28] Biologically active human interferons have also been produced using BPV-based vectors in C127 cells.[29-31]

The main advantage of using BPV expression vectors as compared to using CHO co-amplification systems is time. In order to get high gene copy number, and therefore high expression levels in CHO cells, it is necessary to select for cells growing in higher and higher levels of the selective agent. This is a time-consuming process. With BPV vectors, relatively high-producing cell lines with high gene copy numbers can be isolated a few days after transfection and transformation. Another possible advantage of BPV-based vectors is their potential to be stably maintained as multicopy, extrachromosomal elements in C127 cells. Each gene to be expressed is flanked by the same, known DNA sequences. Heterogeneity in expression levels can result when the genes to be expressed become integrated into chromosomal DNA. By remaining extrachromosomal, the BPV-vectors might be expected to give more reproducible results, free from chromosomal position effects.[30]

A fundamental consideration in the use of BPV vector/C127 cell expression systems is that the cells are anchorage-dependent. This may present problems for large-scale production. BPV vectors have been introduced into a variety of host cells but, at present, all the best host cell lines are anchorage-dependent (see Ref. 5). Future work may identify BPV/host cell lines capable of growing in suspension.

III. Murine Myeloma Cell Lines

Murine myeloma cell lines have two strong points recommending them as the starting point for developing a good mammalian cell expression system. Firstly, they are capable of producing very high levels of a particular protein from single-copy genes. Hybridomas can produce monoclonal antibodies at $130\,\mu g/10^6$ cells/day.[32] Secondly, successful methods for large-scale growth and production of monoclonal antibodies from murine hybridoma cells have already been developed.[33] Presumably it should be possible to produce a recombinant DNA protein in a murine myeloma cell line at levels and costs equal to that of producing monoclonal antibodies. In practice, genetically-engineered antibody genes re-introduced into non-producing myeloma cell lines have produced, at best, levels equal to a low-level producing hybridoma cell line (see Table 1). Non-producing myeloma cell lines, such as Sp2/OAg14, have also been tried for producing non-immunoglobulin recombinant DNA proteins, such as tPA. The myeloma cell lines did secrete tPA, but at low levels ($1\,\mu g/10^6$ cells/day). Both mouse immunoglobulin promoters plus enhancers, and SV40 early promoter plus enhancer, were tried. The SV40 early promoter, plus SV40 enhancer, was 3–10 times better than the immunoglobulin promoters and enhancers.[34]

Making expression vectors that employ the mouse immunoglobulin promoter and enhancer and introducing these vectors into mouse myeloma cell lines does not appear to recreate the high expression levels seen for the endogenous mouse antibody genes making monoclonal antibodies (see Table 1 and Ref. 34). One explanation for this is that the recombinant genes introduced into the myeloma cell lines integrate randomly into mouse chromosomal DNA. They may not, therefore, be in the correct chromosomal environment for high level expression. When genes are introduced into cultured cell lines, it is common to see variable levels of expression of the introduced gene that are not correlated to copy number. The level of expression appears to be influenced by the position of integration. Position effect may occur because the introduced gene does not have sufficient flanking sequence to function independently of the site of integration. Recent studies of the expression and regulation of the human β-globin locus have identified DNA regions flanking the human β-globin locus that contain dominant regulatory sequences that specify position-independent expression. When human β-globin gene flanked by these regulatory regions is introduced into mice, the transgenic mice tissue specifically express the human β-globin gene at high levels, equal to the level of expression of the endogenous mouse β-globin gene. The expression levels are directly related to copy number and independent of position of integration[35] (see Chapter 10, this volume).

The implication of the work with the β-globin gene locus is that other genes expressed at high levels in specific tissues at specific times will have similar regulatory regions. For example, the high levels of expression of the antibody genes in hybridoma cell lines may be due to similar type regulatory regions in the immunoglobulin gene locus. When recombinant genes are transfected into myeloma cell lines, they lack these immunoglobulin regulatory regions and are, therefore, not generally expressed at high levels and very susceptible to position effects. Much better vectors for expressing recombinant genes in myeloma cell lines could be developed if dominant control regions, similar to those of the human β-globin locus, could be identified for the immunoglobulin locus. Inclusion of the proposed dominant immunoglobulin locus control regions in the expression vector may enable a single-copy gene to be expressed at high levels equal to the levels of expression of an endogenous mouse antibody gene. Such studies are underway and are among the most promising new approaches to developing better mammalian cell expression systems.

THE IDEAL EXPRESSION SYSTEM VERSUS THE PRESENT SYSTEMS

In developing new and better mammalian cell expression systems it is important to design expression systems that are compatible with economic large-scale production systems. Table 6 gives a list of

Table 6
Characteristics of the Ideal Mammalian Cell Expression System

1. High production rate per cell
2. Cell lines that give stable production levels over extended storage and cultivation periods
3. Production in serum-free medium
4. Production in the absence of cell growth
5. Production in the absence of toxic selective agents
6. Expression vectors that give high messenger RNA levels and, therefore, high production levels from a single gene copy
7. Expression vectors that do not require extended selection times to increase gene copy number
8. Safe cell lines, not immunogenic, not carcinogenic, not harbouring viruses

characteristics of the ideal mammalian cell expression system keeping suitability for large-scale production in mind. From the point of view of those responsible for process development and scale-up, the goals of mammalian expression technology should be: high rates of foreign protein production, production in the absence of cell growth, high production rates from single-copy genes, stable cell lines, no requirement for toxic selective agents, little or no requirement for serum or serum substitutes, and general safety.[36] This ideal mammalian cell expression system does not yet exist. This list does serve as a means of comparing various expression systems and helps to point out their strengths and weaknesses.

Comparing production levels leads to the conclusion that, at present, the highest levels of protein production are achieved using amplifiable vectors in CHO cells. In CHO cells, production levels of $3-6 \times 10^{14}$ molecules/10^6 cells/day are reported by different groups for a variety of proteins (see Table 4). Although equally high production levels have been achieved using BPV vectors in C127 cells,[25] the more representative figures for good production from BPV/C127 cells fall in the range of $10^{12}-10^{13}$ molecules/10^6 cells/day (see Table 5 and review by Stephens and Hentschel[5]).

The gene coding for HBsAg has been expressed by many groups using a variety of vectors in different cell types, and thus provides a common basis for comparing mammalian cell expression systems (see Table 7). DHFR gene amplification vectors in CHO cells give the highest levels of HBsAg production producing 15 μg/10^6 cells/day.[16] BPV vectors in C127 cells produce up to 1 μg/10^6 cells/day.[25] As discussed earlier, the CHO cell lines producing 15 μg of HBsAg are not stable in the absence of MTX. CHO/DHFR cell lines resistant to lower MTX concentrations and having less highly amplified gene copy numbers are stable in the absence of MTX but produce only 1·5 μg/10^6 cells/day of HBsAg.[16] The C127/BPV cell line producing 1 μg/10^6 cell/day does not require a toxic, selective agent in the medium and could be maintained in continuous production for 60 days.[26] Other mammalian cell expression systems can sometimes produce as much, or nearly as much, HBsAg as the best C127/BPV system (see Table 4). For example, NIH 3T3 mouse fibroblast cells produced 1 μg/10^6 cells/day of HBsAg when the gene copy number was amplified by using a DHFR vector and selecting for resistance to MTX.[13] A Vero cell line has also been reported to produce 1 μg/10^6 cells/day of HBsAg. Vero cells are diploid African green monkey cells that have been approved by the World Health Organization for use in vaccine production. The stable relatively high HBsAg-producing Vero cell line, however, was a rare finding.[37]

Table 7

Examples of Mammalian Cell Expression Systems Used to Produce Hepatitis B Surface Antigen

Cell	Vector	Production levels		References
		µg/10⁶ cells/day	Molecules/10⁶ cells/day	
CHO	DHFR selection, no amplification, HBsAg promoter	0.2^a	4.4×10^{12}	19
CHO	DHFR selection and amplification, HBsAg promoter	15^b	3.3×10^{14}	16
C127	BPV vector, HBsAg promoter	0.2	4.4×10^{12}	55
C127	BPV vector, HBsAg promoter	0.5	1.1×10^{12}	56
C127	BPV vector, MMT-I promoter	1.0	2.2×10^{13}	26
NIH 3T3	BPV vector, HBsAg promoter	0.6	1.3×10^{13}	57
NIH 3T3	MSV^c vector, HBsAg promoter	0.45	9.9×10^{12}	58
NIH 3T3	DHFR selection and amplification, HBsAg promoter	1.0	2.2×10^{13}	13
L tk⁻	tk⁺ vector^d, HBsAg promoter	0.2	4.4×10^{12}	59
L tk⁻	tk⁺ vector, HBsAg promoter	0.15	3.3×10^{12}	60
L tk⁻	tk⁺ vector, HBsAg or SV40 early promoter	0.6	1.3×10^{13}	61
L tk⁻	tk⁺ vector, SV40 early promoter	0.43	9.5×10^{12}	62
Vero	*neo* vector^e HBsAg promoter	1.0^f	2.2×10^{13}	36

[a] Assumes approximately 10⁶ cells/ml.
[b] Not stable, cell lines producing 1·5 µg/10⁶ cells/day were stable.
[c] Moloney mouse sarcoma virus.
[d] Herpes simplex virus thymidine kinase gene for selection.
[e] Confers resistance to aminoglycoside G-418 for selection.
[f] Rare stable cell line.

In addition to comparing the various mammalian expression systems on the basis of best production levels and stability, it is important to consider their growth properties. CHO can be grown in suspension; whereas, C127 cells and NIH 3T3 cells are anchorage-dependent. In general, suspension cells are preferred in production processes.[38]

SUMMARY

It is now possible to produce proteins for scientific, therapeutic and commercial purposes by expressing cloned genes in mammalian cells. Although a fairly wide variety of expression vectors have been developed for a more limited number of mammalian cell types, two mammalian cell expression systems are currently the most successful and widely-used for preparing recombinant DNA products. Both systems rely on having many copies of the recombinant gene per cell in order to obtain high levels of recombinant protein production. The best all-round mammalian cell expression system is based on using DHFR gene-amplification vectors in CHO cells. Another successful, but perhaps less versatile, system is based on using BPV vectors in C127 cells. A serious disadvantage of these two systems is that cells containing high copy numbers of some foreign gene tend to be unstable.

The ideal expression system would give high levels of foreign protein production from a single gene copy. In addition, the system would employ a mammalian cell type that was easy and economical to grow in large-scale cultures. Certain mammalian cell types are capable naturally of producing high levels of a single protein from single copy genes, for example, haemoglobin-producing pre-erythrocytes and immunoglobulin-producing lymphocytes. It has been widely demonstrated that hybridoma cell lines can produce high levels of monoclonal antibodies from single gene copies and that these cells can be grown successfully in large-scale cultures. There are two approaches to developing better mammalian cell expression systems based on using myeloma-type cell lines and immunoglobulin gene regulatory signals. One is identifying and cloning the regulatory sequences that are responsible for high-level expression from single copies of the expressed immunoglobulin genes. These immunoglobulin-regulatory sequences would then be used in constructing expression vectors. Another possibility is targeted homologous recombination; whereby, the recombinant gene is specifically introduced into the genome in the very active immunoglobulin gene locus.[39]

In conclusion, although it is possible to produce foreign protein in mammalian cells, the production efficiencies are still relatively low and the costs relatively high. In practice, this means that only proteins with a high commercial value, generally those with human diagnostic or therapeutic applications, are commercially produced in mammalian cells. By understanding how some mammalian cells can naturally produce large amounts of a particular protein, we may be able to develop better mammalian cell expression systems.

REFERENCES

1. Schnee, J. M., Runge, M. S., Matsueda, G. R., Hudson, N. W., Seidman, J. G., Haber, E. & Quertermous, T., Construction and expression of a recombinant antibody-targeted plasminogen activator. *Proc. Nat. Acad. Sci., USA*, **84** (1987) 6904–8.
2. Harris, T. J. R., Expression of eukaryotic genes in *E. coli*. In *Genetic Engineering*, Vol. 4, ed. R. Williamson, Academic Press, London, 1983, pp. 127–85.
3. Bebbington, C. R. & Hentschel, C. C., The use of vectors based on gene-amplification for the expression of cloned genes in mammalian cells. In *DNA Cloning, Vol. III*, ed. D. M. Glover. IRL Press, Oxford, 1987, pp. 163–88.
4. Kaufman, R. J. & Sharp, P. A., Amplification and expression of sequences cotransfected with a modular dihydrofolate reductase complementary DNA gene. *J. Molec. Biol.*, **159** (1982) 601–21.
5. Stephens, P. E. & Hentschel, C. C., The bovine papilloma virus genome and its uses as a eukaryotic vector. *Biochem. J.*, **248** (1987) 1–11.
6. Mulligan, R. C. & Berg, P., Selection for animal cells that express the *Escherichia coli* gene coding for xanthine-guanine phosphoribosyltransferase. *Proc. Nat. Acad. Sci., USA*, **78**, 2072–6.
7. Morrison, S. L., Transfectomas provide novel chimeric antibodies. *Science*, **229** (1985) 1202–7.
8. Swartz, R., Selecting a preparative mammalian cell production technology in 1988. *Gen. Engng News*, **8** (1988) 14–35.
9. Schimke, R. T., Gene amplification in cultured animal cells. *Cell*, **37** (1984) 705–13.
10. Stark, G. R. & Wahl, G. M., Gene amplification. *Ann. Rev. Biochem.*, **53** (1984) 447–91.
11. Stark, G. R., DNA amplification in drug resistant cells and in tumours. *Cancer Surveys*, **5** (1986) 1–23.
12. Kaufman, R. J., Murtha, P., Ingolia, D. E., Yeung, C. Y. & Kellens, R. E., Selection and amplification of heterologous genes encoding adenosine deaminase in mammalian cells. *Proc. Nat. Acad. Sci., USA*, **83** (1986) 3136–40.
13. Christman, J. K., Gerber, M., Price, P. M., Flordellis, C., Edelman, J. & Acs, G., Amplification of expression of hepatitis B surface anitgen in 3T3 cells cotransfected with a dominant-acting gene and cloned viral DNA. *Proc. Nat. Acad. Sci., USA*, **79** (1982) 1815–19.
14. Murray, M. J., Kaufman, R. J., Latt, S. A. & Weinberg, R. A., Construction and use of a dominant, selectable marker: a Harvey sarcoma virus-dihydrofolate reductase chimera. *Molec. Cell Biol.*, **3** (1983) 32–43.
15. Deschatrette, J., Fougere-Deschatrette, C., Corcos, L. & Schimke, R. T., Expression of the mouse serum albumin gene introduced into differentiated

and dedifferentiated rat hepatoma cells. *Proc. Nat. Acad. Sci., USA,* **82** (1985) 765–9.

16. Michel, M. L., Sobczak, E., Malpiece, Y., Tiollais, P. & Streeck, R. E., Expression of amplified hepatitis B virus surface antigen genes in Chinese hamster ovary cells. *Bio/Technology,* **3** (1985) 561–6.

17. Kaufman, R. J., Wasley, L. C., Spiliotes, A. J., Gossels, S. D., Latt, S. A., Larsen, G. R. & Kay, R. M., Coamplification and coexpression of human tissue-type plasminogen activator and murine dihydrofolate reductase sequences in Chinese hamster ovary cells. *Molec. Cell. Biol.,* **5** (1985) 1750–9.

18. Mory, Y., Ben-Barak, J., Seger, D., Cohen, B., Novick, D., Fischer, D. G., Rubinstein, M., Kargman, S., Zilberstein, A., Vigneron, M. & Revel, M., Efficient constitutive production of human IFN-γ in Chinese hamster ovary cells. *DNA,* **5** (1986) 181–93.

19. Patzer, E. J., Nakamura, G. R., Hershberg, R. D., Gregory, T. J., Crowley, C., Levinson, A. D. & Eichberg, J. W., Cell culture derived recombinant HBsAg is highly immunogenic and protects chimpanzees from infection with hepatitis B virus. *Bio/Technology,* **4** (1986) 630–6.

20. DiMaio, D., Papillomavirus cloning vectors. In *The Papillomaviruses,* Plenum Press, New York, 1984.

21. Lowy, D. R., Rands, E. & Scolnick, E. M., Helper-independent transformation by unintegrated Harvey sarcoma virus DNA. *J. Virol.,* **26** (1978) 291–8.

22. Lowy, D. R., Dvoretzky, I., Shober, R., Law, M. F., Engel, L. & Howley, P. M., In vitro transformation by a defined subgenomic fragment of bovine papilloma virus DNA. *Nature,* **287** (1980) 72–4.

23. Campo, M. S., Spandidos, D. A., Lang, J. and Wilkie, N. M., Transcriptional control signals in the genome of bovine papillomavirus type 1. *Nature,* **303** (1983) 77–80.

24. Sambrook, J., Rodgers, L., White, J. & Gething, M. J., Lines of BPV-transformed murine cells that constitutively express influenza virus hemagglutinin. *EMBO J.,* **4** (1985) 91–103.

25. Pavlakis, G. N. & Hamer, D. H., Regulation of a metallothionein-growth hormone hybrid gene in bovine papilloma virus. *Proc. Nat. Acad. Sci., USA,* **80** (1983) 397.

26. Hsiung, N., Fitts, R., Wilson, S., Milne, A. & Hamer, D., Efficient production of hepatitis B surface antigen using a bovine papilloma virus-metallothienein vector. *J. Molec. Appl. Genet.,* **2** (1984) 497–506.

27. Stephens, P. E., Bendig, M. M. & Hentschel, C. C., Expression of human tissue plasminogen activator in mouse cells. *J. Cell Biochem.,* **Suppl. 9C** (1985) 75.

28. Bendig, M. M., Stephens, P. E., Cockett, M. I. & Hentschel, C. C., Mouse cell lines that use heat shock promoters to regulate the expression of tissue plasminogen activator. *DNA,* **6** (1987) 343–52.

29. Fukunaga, R., Sokawa, Y. & Nagata, S., Constitutive production of human interferons by mouse cells with bovine papillomavirus as a vector. *Proc. Nat. Acad. Sci., USA,* **81** (1984) 5086–90.

30. Zinn, K., Mellon, P., Ptashne, M. & Maniatis, T., Regulated expression of an extrachromosomal human β-interferon gene in mouse cells. *Proc. Nat. Acad. Sci., USA,* **79** (1982) 4897–901.

31. Mitrani-Rosenbaum, S., Maroteaux, L., Mory, Y., Revel, M. & Howley, P. M.,

Inducible expression of the human interferon 1 gene linked to a bovine papilloma virus DNA vector and maintained extrachromosomally in mouse cells. *Molec. Cell Biol.,* **3** (1983) 233–240.

32. de St. Groth, S. F., Automated production of monoclonal antibodies in a cytostat. *J. Immunol. Methods,* **57** (1983) 121–36.

33. Birch, J. R., Boraston, R. & Wood, L., Bulk production of monoclonal antibodies in fermenters. *Trends Biotechnol.,* **3** (1985) 162–6.

34. Weidle, U. H. & Buckel, P., Establishment of stable mouse myeloma cells constitutively secreting human tissue-type plasminogen activator. *Gene,* **57** (1987) 131–41.

35. Grosveld, F., van Assendelft, G. B., Greaves, D. R. & Kollias, G., Position-independent, high-level expression of the human β-globin gene in transgenic mice. *Cell,* **51** (1987) 975–85.

36. Swartz, R., Alternatives for the production of mammalian cell products: A survey. *Gen. Engng News,* **5** (1985) 16–21.

37. Colbere-Garapin, F., Horaud, F., Kourilsky, P. & Garapin, A. C., Stable HBV surface antigen expression by Vero cell clones after transfection. *Develop. Biol. Standard.,* **59** (1985) 109–12.

38. Arathoon, W. R. & Birch, J. R., Large-scale cell culture in biotechnology. *Science,* **232** (1986) 1390–5.

39. Sedivy, J. M., New genetic methods for mammalian cells. *Bio/Technology,* **6** (1988) 1192–6.

40. Morrison, S. L., Johnson, M. J., Herzenberg, L. A. & Oi, V. T., Chimeric human antibody molecules: Mouse antigen-binding domains with human constant region domains. *Proc. Nat. Acad. Sci., USA,* **81** (1984) 6851–5.

41. Boulianne, G. L., Hozumi, N. & Shulman, M. J., Production of functional chimeric mouse/human antibody. *Nature,* **312** (1984) 643–6.

42. Boulianne, G. L., Iseman, D. E., Hozumi, N. and Shulman, M. J., Biological properties of chimeric antibodies: Interaction with complement. *Molec. Biol. Med.,* **4** (1987) 37–49.

43. Sahagan, B. G., Dorai, H., Saltzgaber-Muller, J., Toneguzzo, F., Guindon, C. A., Lilly, S. P., McDonald, K. W., Morrissey, D. V., Stone, B. A., Davis, G. L., McIntosh, P. K. & Moore, G. P., A genetically engineered murine/human chimeric antibody retains specificity for human tumor-associated antigen. *J. Immunol.,* **137** (1986) 1066–74.

44. Sun, L. K., Curtis, P., Rakowicz-Szulczynska, E., Ghrayeb, J., Chang, N., Morrison, S. L. & Koprowski, H., Chimeric antibody with human constant regions and mouse variable regions directed against carcinoma-associated antigen 17-1A. *Proc. Nat. Acad. Sci., USA,* **84** (1987) 214–18.

45. Steplewski, Z., Sun, L. K., Shearman, C. W., Ghrayeb, J., Daddona, P. & Koprowski, H., Biological activity of human–mouse IgG1, IgG2, IgG3, and IgG4 chimeric monoclonal antibodies with antitumor specificity. *Proc. Nat. Acad. Sci., USA,* **85** (1988) 4852–6.

46. Nishimura, Y., Yokoyama, M., Araki, K., Veda, R., Kudo, A. & Watanabe, T., Recombinant human–mouse chimeric monoclonal antibody specific for common acute lymphocytic leukemia antigen. *Cancer Res.,* **47** (1987) 999–1005.

47. Liu, A. Y., Robinson, R. R., Hellstrom, K. E., Murray, E. D., Chang, C. P. & Hellstrom, I., Chimeric mouse–human IgG1 antibody that can mediate lysis of cancer cells. *Proc. Nat. Acad. Sci., USA,* **84** (1987) 3439–43.

48. Liu, A. Y., Mack, P. W., Champion, C. I. & Robinson, R. R., Expression of mouse:human immunoglobulin heavy-chain cDNA in lymphoid cells. *Gene,* **54** (1987) 33–40.
49. Riechmann, L., Clark, M., Waldmann, H. & Winter, G., Reshaping human antibodies for therapy. *Nature,* **332** (1988) 323–7.
50. Falkner, F. G. & Zachau, H. G., Expression of mouse immunoglobulin genes in monkey cells. *Nature,* **298** (1982) 286–8.
51. Weidle, U. H., Borgya, A., Mattes, R., Lenz, H. & Buckel, P., Reconstitution of functionally active antibody directed against creatine kinase from separately expressed heavy and light chains in non-lymphoid cells. *Gene,* **51** (1987) 21–9.
52. Bodmer, M., In vitro and in vivo activities of a mouse–human chimeric B72.3 antibody. Presented at Advances in the applications of monoclonal antibodies in clinical oncology, University of London Royal Postgraduate Medical School, 25–27 May, 1988.
53. Vehar, G. A., Kohr, W. J., Bennett, W. F., Pennica, D., Ward, C. A., Harkins, R. N. & Collen, D., Characterization studies on human melanoma cell tissue plasminogen activator. *Bio/Technology,* **2** (1984) 1051–7.
54. Ramabhadran, T. V., Reitz, B. A. & Tiemeier, D. C., Synthesis and glycosylation of the common subunit of human glycoprotein hormones in mouse cells. *Proc. Nat. Acad. Sci., USA,* **81** (1984) 6701–5.
55. Stenlund, A., Lamy, D., Moreno-Lopez, J., Ahola, H., Petterson, U. & Tiollais, P., Secretion of the hepatitis B virus surface antigen from mouse cells using an extra-chromosomal eucaryotic vector. *EMBO J.,* **2** (1983) 669–73.
56. Denniston, K. J., Yoneyama, T., Hoyer, B. H. & Gerin, J. L., Expression of hepatitis B virus surface and e antigen genes cloned in bovine papilloma vectors. *Gene,* **32** (1984) 357–68.
57. Wang, Y., Stratowa, C., Schaefer-Ridder, M., Doehmer, J. & Hofschneider, P. H., Enhanced production of hepatitis B surface antigen in NIH 3T3 mouse fibroblasts by using extrachromosomally replicating bovine papillomavirus vector. *Molec. Cell Biol.,* **3** (1983) 1032–9.
58. Stratowa, C., Doehmer, J., Wang, Y. & Hofschneider, P. H., Recombinant retroviral DNA yielding high expression of hepatitis B surface antigen. *EMBO J.,* **1** (1982) 1573–8.
59. Dubois, M. F., Pourcel, C., Rousset, S., Chany, C. & Tiollais, P., Excretion of hepatitis B surface antigen particles from mouse cells transformed with cloned viral DNA. *Proc. Nat. Acad. Sci., USA,* **77** (1980) 4549–53.
60. Gough, N. M. & Murray, K., Expression of the hepatitis B virus surface, core and e antigen genes by stable rat and mouse cell lines. *J. Molec. Biol.,* **162** (1982) 43–67.
61. Malpiece, Y., Michael, M. L., Carloni, G., Revel, M., Tiollais, P. & Weissenbach, J., The gene S promoter of hepatitis B virus confers constitutive gene expression. *Nucl. Acids Res.,* **11** (1983) 4645–54.
62. Carloni, G., Malpiece, Y., Michel, M. L., Le Patezour, A., Sobczak, E., Tiollais, P. & Streeck, R. E., A transformed Vero cell line stably producing the hepatitis B virus surface antigen. *Gene,* **31** (1984) 49–57.

Chapter 7

EXPRESSION OF FOREIGN PROTEINS BY VACCINIA VIRUS

GEOFFREY L. SMITH*

Department of Pathology, University of Cambridge, Cambridge, UK

INTRODUCTION

In 1801, 3 years after he had introduced vaccination,[1] Edward Jenner made his famous prediction, '. . . that the annihilation of the smallpox, the most dreaded scourge of the human species, must be the final result of this practice'. One hundred and seventy six years later this prophecy was fulfilled.[2] This triumph of preventive medicine was achieved by using the antigenically related orthopoxvirus, vaccinia, for immuno-prophylaxis against variola virus the causative agent of smallpox. It might also have been predicted that once smallpox had been eradicated interest in poxviruses would diminish. Paradoxically research on vaccinia and other poxviruses is now more intense than ever. A major factor contributing to this increased interest is the development of techniques that permit the expression of foreign proteins from re-combinant poxviruses. These expression vectors have a variety of applications which include potential as new live vaccines in veterinary or human medicine. The construction and application of recombinant vaccinia viruses is the topic of this chapter.

VACCINIA MOLECULAR BIOLOGY

Vaccinia virus is a member of the family poxviridae and genus orthopoxvirus. Characteristics of these viruses are a large and complex particle morphology, large double stranded DNA genomes and cyto-plasmic replication.[3] Much of our knowledge of poxvirus molecular

*Present address: Sir William Dunn School of Pathology, Oxford University, Oxford, UK.

biology is based on studies with vaccinia, the prototype orthopoxvirus, but an increasing number of characteristics of vaccinia virus are proving common to other poxviruses.

Virus Particle

The virus particle is roughly 200 nm by 300 nm and is brick-shaped with irregularly arranged surface tubules covering the outer layer of the lipid envelope. Inside the envelope there are two lateral bodies associated with a biconcave core. There is no icosahedral symmetry. The core contains the DNA genome, associated proteins and many enzymes. Virus particles released from the cell contain an additional lipid envelope and associated glycoproteins.[4] Both forms of particle are infectious.

Virus Genome

The genome is a double stranded DNA molecule of 185 kb. At its termini the two DNA strands are linked by covalently closed hairpin loops.[5] Proximal to the terminal hairpins are large inverted repeats (ITR) of 10 kb. There are also short tandemly repeated sequences of unknown function within the ITRs. The remaining 165 kb of the genome may be divided into three regions. Regions extending inward for up to 15 kb from the left or right inverted terminal repeat are poorly conserved among different strains of vaccinia or other orthopoxviruses[6] and are non-essential for replication in most tissue culture systems. The remaining central region is more highly conserved and contains many genes essential for virus replication. The lesions of most conditional lethal mutants map in this conserved region.[7] Virus genes are tightly packed with little intervening non-coding regions and with occasional short overlaps. The nucleotide sequence of greater than half the genome has been determined.

Virus Proteins

Highly purified intracellular virus particles contain greater than 100 polypeptides. These exhibit a wide range of sizes (10–140 kD) and there are several types of post-translational modification. A 58 kD protein forming the surface tubules and a 14 kD envelope protein are targets for neutralising antibodies. Extracellular virus particles possess additional glycoproteins including the haemagglutinin that contains N- and O-linked sugar residues. Within infected cells greater than 200 virus induced polypeptides are expressed in a temporally and quantitatively regulated manner.

A feature of poxviruses is the expression of many virus-coded enzymes. Table 1 lists those enzymes present in virus particles and additional enzymes expressed in infected cells. Many of these enzymes are required to enable this virus to replicate within the cytoplasm. The capping and methylating enzymes have proved useful to the molecular biologist.[8]

Virus Gene Expression

(i) Early

Within minutes of virus penetration into the cytoplasm early virus mRNAs are synthesised by a process that is independent of host protein and nucleic acid synthesis. These transcripts contain a 5′ methylated cap, 3′ poly A and are of discrete length. The early mRNAs cover approximately 50% of the virus genome and code up to 100 genes that are distributed throughout the genome on both DNA strands.

Early transcription is performed by the virus-coded RNA polymerase that is present within the virus particles.[9] This enzyme is a multimeric complex (M_r 500 000 D) with similarities to host eukaryotic RNA polymerase II in terms of the size and number of constituent subunits.[10] There is also limited amino acid homology[11] and immunological cross-reactivity.[12] Despite these similarities the vaccinia RNA polymerase is

Table 1
Vaccinia Virus Enzymes

A. Virus particle
 DNA-dependent RNA polymerase
 RNA guanyltransferase
 RNA (guanine-7-) methyltransferase
 RNA (nucleoside-2-) methyltransferase
 Poly A polymerase
 Endoribonuclease
 5′ phosphate polynucleotide kinase
 5′ RNA triphosphatase
 DNA-dependent ATPase
 DNA/RNA-dependent NTPase
 Deoxyribonuclease
 DNA topoisomerase
 Protein kinase
 Alkaline protease

B. Infected cell
 Thymidine kinase
 Ribonucleotide reductase
 DNA polymerase
 DNA ligase

functionally distinct and will only transcribe genes from poxviruses[13] or other cytoplasmically replicating DNA viruses such as African Swine Fever Virus. Conversely, the host RNA polymerase II will transcribe host but not poxvirus genes. In accord with these observations, comparative analyses of early vaccinia promoters showed these to be structurally distinct from host RNA polymerase II promoters. In-vitro mutagenesis demonstrated that the region up to −30 from the RNA start site contains all the signals necessary for early transcription.[14-16] This region is very A : T rich. Early transcription factors have been identified and include a 130-kD virion component with 'footprints' −15 to −29 from the RNA start site,[17] and a heterodimeric DNA-dependent ATPase.[18] Termination of early transcription occurs approximately 50 nucleotides downstream of a motif TTTTTNT.[19] This signal is recognised in the RNA form (UUUUUNU) by a termination factor which is also the RNA capping enzyme complex.[20] Early genes code for many enzymes involved in nucleic acid metabolism such as RNA and DNA polymerase, thymidine kinase and ribonucleotide reductase. Another interesting early gene product is a vaccinia growth factor.[21]

(ii) Late

Late genes are transcribed after the onset of virus DNA replication. The late mRNAs are large and heterogeneous in length, possess a 5′ methylated cap and, remarkably, contain a poly A tract at both the 5′ and 3′ ends.[22-24] The size heterogeneity results from failure to terminate transcription at specific sites, the TTTTTNT early signal no longer being recognised. The 5′ poly A tract is approximately 35 residues long, is not present in template DNA, and has been proposed to arise by transcriptional slippage during copying of a short series of thymidine residues.[25] Its functional role and influence on translation remain to be determined.

Late promoters contain a conserved motif TAAAT(G).[26] The 3′ nucleotide is often G and completes a translation initiation codon. S1 nuclease mapping of the 5′ end of the RNA indicate that transcription initiates within the three A residues of this motif. Sequences of only 15–20 nucleotides upstream from the RNA start site function as the late promoters.[27, 28]

Possible mechanisms causing the switch from early to late gene transcription are modification of the RNA polymerase complex or alteration in structure of the replicating DNA template.

Late genes code for many of the major structural components of the virus particle.

Virus Morphogenesis

The maturation of vaccinia virus particles is unusual in several respects. Lipid crescents, which ultimately become the virus envelope, are synthesised *de novo* within the cytoplasm and this process may be reversibly blocked by the drug rifampicin.[29] The majority of infectious progeny remains cell-associated but a small fraction is released by a process resulting in acquisition of a second lipid envelope. Virus particles that are to egress from the cell become wrapped by a double layer of membrane derived from the Golgi apparatus. One of these extra membranes is then lost by fusion at the cell surface. The glycoproteins that are present on only the extracellular virus give these particles different biological and antigenic characteristics.

RECOMBINANT VACCINIA VIRUSES

Construction

The vaccinia genome is too large to be conveniently manipulated *in vitro* and is also non-infectious without associated transcriptional enzymes. Consequently, recombinant viruses are constructed by manipulation of cloned virus restriction fragments followed by homologous recombination after introduction of these DNA fragments into virus-infected cells.[30-33] Plasmids called insertion vectors have been constructed to facilitate the rapid construction of vaccinia virus recombinants (Fig. 1). These

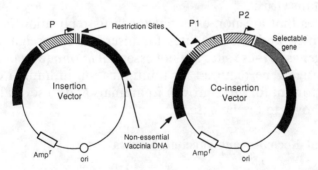

Fig. 1. Plasmid vectors for inserting genes into vaccinia virus. The insertion vector contains a single vaccinia promoter (P) upstream of unique restriction fragments. The flanking DNA is taken from a non-essential region of the virus genome such as TK. The co-insertion vector contains similar features plus a second vaccinia promoter driving expression of a gene such as β-galactosidase or Ecogpt that aids selection of the recombinant virus.

plasmids contain a strong poxvirus promoter and downstream restriction sites for insertion of a foreign gene flanked by vaccinia DNA taken from a non-essential region of the genome. Such a plasmid, when transfected into vaccinia infected cells, can recombine with the homologous sequences of the virus genome resulting in the insertion of the foreign gene. Recombinant genomes are then replicated and packaged into infectious progeny virus. The temporal and quantitative control of the foreign gene are determined by the choice of virus promoter and the site of insertion into the virus genome is determined by the flanking non-essential DNA.

Between 0·1% and 1·0% of progeny virus from transfected cells are recombinant. Therefore, a variety of selective methods have been employed to distinguish recombinant from parental virus. DNA hybridisation with a probe specific for the foreign DNA allows identification of recombinant plaques.[32,33] Genetic selection has been more widely used. Insertion of a foreign gene into the endogenous vaccinia thymidine kinase (TK) gene permits selection of TK⁻ viruses using bromodeoxyuridine (BudR).[30,31] Alternatively, insertion of the herpes simplex virus TK gene into vaccinia virus TK⁻ mutants has permitted the selection of TK⁺ recombinants.[30,32] More recently, other dominant selectable markers, neomycin resistance[34] and *E. coli* xanthine guanine phosphoribosyltransferase (Ecogpt)[35,36] have enabled selection of recombinants without use of special cell lines. Expression of β-galactosidase[37] or luciferase[38] allows visual detection of recombinant plaques. Another approach has been the use of conditional lethal mutants and selection of recombinants under conditions non-permissive for the mutant virus.[39,40]

Many sites that are non-essential for virus replication *in vitro* have been identified and used as sites for insertion of foreign DNA.[30,41,42] While the genes at these sites are not essential *in vitro* (except in the case of a host range gene required for multiplication in human cells), they may be important for virus growth in animals and in several cases this has been shown to be the case.[42-44]

Properties of Recombinant Vaccinia Viruses

Recombinant vaccinia viruses have a large capacity for foreign DNA,[45] have a broad host range in cell culture and in animals and are infectious. They can be easily grown in cell culture to titres in excess of 10^{10} pfu/ml. The genome of the recombinant virus will stably contain and express the foreign gene unless (a) the foreign protein is toxic to virus replication; (b) there are tandemly repeated copies of the virus promoter or foreign gene; or (c) the gene is inserted into inherently unstable regions of the genome such as adjacent to the ITRs. Examples of toxic foreign proteins are lacking but such problems might be overcome by using inducible promoter systems. Tandemly repeated genes will resolve via homologous recombination to yield a single stable gene. Insertion of foreign genes into variable regions of the virus genome might rarely result in spontaneous deletion if the virus is grown in cell culture. When such regions of the virus genome are used careful analysis of the recombinant genome by Southern blotting is especially important.

Gene Expression

Genes from diverse origins have been expressed from vaccinia virus. The major requirements are that the gene should be a contiguous protein coding sequence without introns, the sequence TTTTTNT should be absent and the first ATG codon downstream of the RNA start site of the vaccinia promoter should be the one initiating translation of the foreign protein. For genes containing introns cDNA copies must be used. The presence of the early transcriptional termination sequence TTTTTNT may be overcome by either placing the gene under a late promoter or by mutation of this motif. Extra upstream initiation codons diminish translation from the authentic start site but may not completely obviate expression. Where possible they should be removed and the length of the 5′ untranslated region reduced to the minimum.

Post-translation modifications of the primary translation product such as proteolysis,[47, 48] glycosylation[49-51] and carboxylation[52] occur and closely match the modifications of the foreign protein in its normal environment. Transport of the protein within the cell or from the cell also occurs normally. It is therefore possible to express authentic eukaryotic proteins in a broad range of eukaryotic cells.

The level and time of expression reflects the promoter chosen. Vaccinia promoters that are expressed early (e.g. TK[15]), late (e.g. 4b[46]) or constitutively (e.g. 7·5 K[14]) have been used. To date the strongest naturally occurring vaccinia promoters come from genes coding for major late structural components of the virion such as the basic 11 K protein[53] or the 4b core polypeptide.[46] β-galactosidase represents 3% of total infected cell protein when expressed from the 11 K promoter.[36] Expression from the widely used 7·5 K promoter yielded 2·6 μg of hepatitis B virus surface antigen per 10^6 cells[49] and in another study using the same promoter roughly one million molecules of influenza haemagglutinin were synthesised per cell.[54] An increase in the level of expression is being approached in three ways. (1) A detailed understanding of the regions of the promoter influencing expression level is being sought by in-vitro mutagenesis of promoters coupled to reporter genes like chloramphenicol acetyl transferase (CAT). These studies and experiments using in-vitro transcription systems derived from virus cores or infected cell extracts are providing detailed information about the essential nucleotides constituting vaccinia promoters and about proteins which interact with these sequences. Construction of synthetic promoters yielding higher levels of expression may then be possible. (2) A hybrid expression system using a vaccinia promoter to express bacteriophage T7 RNA polymerase from one virus and a T7 promoter to express the desired foreign gene from a second virus has been developed.[55, 56] Coinfections with these two viruses yield greater levels of foreign gene expression than obtained with conventional single promoter systems. However, coexpression of the T7 RNA polymerase and the foreign gene driven by a T7 promoter from the same virus was not possible, probably because the high level of T7 transcription was toxic to vaccinia virus replication. (3) Cowpox virus contains an A type cytoplasmic inclusion body which is composed of a very abundant single protein. The gene encoding this protein has been mapped[57] and foreign proteins expressed from its promoter are visible as major bands on SDS-polyacrylamide gels (D. Pickup, personal communication). This may provide levels of expression approaching those obtained from the baculovirus insect cell vector system.

Applications of Recombinant Vaccinia Viruses

The infectivity and broad host range of vaccinia recombinants give this vector system many applications. A use already mentioned is the study of poxvirus gene regulation. It is straightforward to reintroduce into vaccinia virus mutated vaccinia promoters joined to a reporter gene and to determine the effect of mutation upon gene expression.

Another use of vaccinia expression vectors is the identification of antigens which have potential as vaccines against pathogenic organisms. This may be divided into the identification of antigens that are recognised by antibodies which neutralise infectivity, and the identification of antigens recognised by cytotoxic T lymphocytes (CTL) on the surface of infected cells. As an example of the former application, large scale nucleotide sequencing of human cytomegalovirus (HCMV) DNA genome identified genes which coded for proteins which had characteristics of glycoproteins, namely amino and carboxy terminal hydrophobic sequences and potential N-linked glycosylation sites. Such proteins are likely to be expressed on the surface of virus particles or infected cells and hence be targets for neutralising antibodies. Two such genes were expressed in vaccinia virus recombinants and the gene products detected using polyvalent or monoclonal antibodies against HCMV.[58, 59] In each case the gene products are glycoproteins that are transported to cellular membranes and which are targets for neutralising antibodies. In one case the glycoprotein (gB) induced neutralising antibodies when the recombinant vaccinia virus was used to vaccinate animals.[58] These proteins are therefore candidates for inclusion in future anti-HCMV vaccines.

The host's cellular immunity is also an important defence mechanism against virus infections particularly in detecting and destroying virus infected cells. Cytotoxic T cells recognise foreign peptides on the cell surface in association with class I molecules of the major histo-compatibility complex (MHC) and lyse such cells. In terms of vaccines, it is therefore important to identify those virus proteins which are broken down into peptides and presented to the class I restricted CTL, and to understand how CTL directed against these protein fragments are best induced. For the identification of such virus proteins, vaccinia has been a very useful vector system. The usefulness stems from two observations. First, a vaccinia recombinant expressing the influenza HA induced CTL directed against HA after vaccination of animals.[60] Second, CTL taken from mice immunised with the homologous influenza virus strain were able to recognise and destroy histocompatible target cells infected with the vaccinia recombinant expressing HA.[60] Vaccinia recombinants can therefore be used to identify virus proteins that are recognised by CTL either by inducing CTL in animals, or by making target cells against which CTL resulting from virus infection can be tested. For example, a collection of ten vaccinia recombinants individually expressing each influenza virus protein has allowed a systematic study of which antigens are recognised.[61-64] Surprisingly, it is the 'internal' virus proteins, such as nucleoprotein (NP), that are the major targets for CTL.[62] Recognition of individual proteins can also vary with MHC haplotype.[61] Similar systematic studies with other viruses, e.g. respiratory syncitial virus[65] have also been performed and in these cases also it is the internal virus antigens that are the major targets.

The use of vaccinia for studying antigen recognition by CTL also yielded a surprising and puzzling observation. Cells infected with vaccinia recombinants expressing influenza HA were recognised by HA-specific CTL if the HA was expressed from an early but not a late promoter.[66] This observation has been extended to influenza NP and for some NP epitopes the blockage in presentation also occurs early during infection.[67] For both HA and NP the blockage in presentation was partially or completely overcome by expressing unstable, rapidly degraded forms of the proteins.[67] Conceivably, vaccinia might be interfering with antigen presentation by preventing normal protein turnover and peptide formation. In support of this notion genes coding for inhibitors of serine proteases have been found in members of several poxvirus genera (avipox, leporipox and orthopox).[68-71] Very recently, a family of serine protease inhibitors (serpins) have also been identified in vaccinia virus[71] (G. L. Smith, unpublished data) and deletion of such genes will permit an experimental determination of whether their gene products interfere with antigen presentation to CTL. If this is true it would be a novel mechanism by which a virus-infected cell evades recognition by host CTL.

Vaccinia is not an expression system that makes enormous amounts of foreign protein such as *E. coli* or yeast which are discussed in other chapters of this volume. Nonetheless, it may have a role in the production of proteins that must be synthesised in specific types of mammalian cells and which have potent biological activity. The expression of factor VIII described elsewhere in this volume is such an example. Increased levels of expressoin may broaden this application.

The transport of proteins within infected cells has also been studied using recombinant vaccinia viruses. The trafficking of individual virus glycoproteins to the apical or basal surface membrane is one example.[72] Another is the study of the subcellular localisation of individual influenza virus proteins in the absence of other influenza gene products.[73] A third is the identification of a signal specifying transport into the nucleolus.[74]

Vaccines

Vaccinia recombinants have potential application as new live vaccines against a wide range of pathogens. This has been demonstrated by the vaccination of experimental animals and induction of immune responses conferring protection against disease caused by heterologous pathogens.[33, 75-83] The advantages of this approach include: the induction of antibody and cell mediated immune responses; the low cost of vaccine manufacture and ease of administration; the vaccine stability without refrigeration; the broad host range of vaccinia permitting veterinary and human application; and the large capacity for foreign DNA so that polyvalent vaccines[84] may be constructed. The principal disadvantage is the rare occurrence of vaccine-related complications.[85]

Considerable research effort is directed towards construction of attenuated virus strains. This involves the identification and deletion of specific genes that are likely to influence virus virulence. The TK, HA and growth factor genes all contribute to virus pathogenicity and their deletion has resulted in attenuation.[42-44] Vaccinia also expresses a protein with homology to a complement control factor.[86] The protein is secreted from the infected cell and may function to interfere with complement mediated cell lysis. It is likely that deletion of this gene would also attenuate the virus. For veterinary application a vaccinia virus lacking the host range gene required for growth in human cells could prevent accidental infection of humans.[87] This may be especially useful in vaccine field trials against rabies using recombinant vaccinia-infected bait to immunise wild animals.[88] Deletion of serpin genes may also attenuate the virus since virus-infected cells may be more efficiently recognised and lysed by CTL. While any of these approaches individually, or in combination, may attenuate the virus they may also diminish immunogenicity. For the virus to be an effective and safe vaccine, a suitable balance between attenuation and immunogenicity are required. When such a balance is found, vaccinia might again be used for mass vaccination. Perhaps then Jenner's prophecy '. . . that the annihilation of smallpox must be the final result of this practice . . .' may be extended to include other infectious diseases.

ACKNOWLEDGEMENTS

The author thanks Anita Hancock for typing and Paco Rodriguez for critical reading of the manuscript.

REFERENCES

1. Jenner, E., An inquiry into the causes and effects of the variolae vaccinae, a disease discovered in some regions of western counties of England, particularly Gloucestershire, and known by the name of cowpox, 1798. (Reprinted by Cassel, London in 1896.)
2. World Health Organization, The global eradication of smallpox. Final report of the global commission for the certification of smallpox eradication, WHO, Geneva, 1980.
3. Moss, B., Replication of poxviruses. In *Virology,* ed. B. N. Fields, R. M. Chanock & B. Roizman. Raven Press, New York, 1985, pp. 658–703.
4. Boulter, E. A. & Appleyard, G., Differences between the extracellular and intracellular forms of poxvirus and their implications. *Prog. Med. Virol.,* **16** (1973) 86–108.
5. Baroudy, B. M., Venkatesan, S. & Moss, B., Structure and replication of

vaccinia virus telomeres. *Cold Spring Harbor Symposia on Quantitative Biology,* Vol. XLVII, 1983, pp. 723–9.

6. Mackett, M. & Archard, L. E., Conservation and variation in the orthopoxvirus genome structure. *J. Gen. Virol.,* **45** (1979) 658–702.

7. Thompson, C. L. & Condit, R. C., Marker rescue mapping of vaccinia virus temperature sensitive mutants using overlapping cosmid clones representing the entire virus genome. *Virology,* **150** (1986) 10–20.

8. Moss, B., End labelling of RNA with capping and methylating enzymes. In *Gene Amplification and Analysis,* Vol. 2, ed. T. S. Papas, M. Rosenberg & J. G. Chirikjian. Elsevier, New York.

9. Kates, J. R. & McAuslan, B., Poxvirus DNA-dependent RNA polymerase. *Proc. Nat. Acad. Sci., USA,* **58** (1969) 134–41.

10. Jones, E. J., Puckett, C. & Moss, B., DNA-dependent RNA polymerase subunits encoded within the vaccinia virus genome. *J. Virol.,* **61** (1987) 1765–71.

11. Broyles, S. & Moss, B., Homology between RNA polymerases of poxviruses, prokaryotes and eukaryotes: nucleotide sequence and transcriptional analysis of vaccinia virus genes encoding 147 kDa and 22 kDa subunits. *Proc. Nat. Acad. Sci., USA,* **83** (1986) 3141–5.

12. Morrison, D. K. & Moyer, R. W., Detection of a subunit of cellular RNA polymerase II within highly purified preparations of RNA polymerase isolated from rabbit poxvirus virions. *Cell,* **44** (1986) 587–96.

13. Cochran, M. A., Mackett, M. & Moss, B., Eukaryotic transient expression system dependent on transcription factors and regulatory DNA sequences of vaccinia virus. *Proc. Nat. Acad. Sci., USA,* **82** (1985) 19–23.

14. Cochran, M. A., Puckett, C. & Moss, B., *In vitro* mutagenesis of the promoter region of a vaccinia virus gene: evidence for tandem early and late regulatory signals. *J. Virol.,* **54** (1985) 30–7.

15. Weir, J. P. & Moss, B., Determination of the promoter region of an early vaccinia virus gene encoding thymidine kinase. *Virology,* **158** (1987) 206–10.

16. Coupar, B. E. H., Boyle, D. B. & Both, G. W., Effect of *in vitro* mutations in a vaccinia virus early promoter region monitored by herpes simplex virus thymidine kinase expression in recombinant vaccinia virus. *J. Gen. Virol.,* **68** (1987) 2299–309.

17. Yuen, L., Davison, A. J. & Moss, B., Early promoter-binding factor from vaccinia virus virions. *Proc. Nat. Acad. Sci., USA,* **84** (1987) 6069–73.

18. Broyles, S. S. & Moss, B., DNA-dependent ATPase activity associated with vaccinia virus early transcription factor. *J. Biol. Chem.,* **263** (1988) 10761–5.

19. Rohrmann, G., Yuen, L. & Moss, B., Transcription of vaccinia virus early genes by enzymes isolated from vaccinia virions terminates at a downstream regulatory sequence. *Cell,* **46** (1986) 1029–35.

20. Schuman, S., Broyles, S. S. & Moss, B., Purification and characterization of a transcription termination factor from vaccinia virions. *J. Biol. Chem.,* **262** (1987) 12372–80.

21. Brown, J. P., Twardzik, D. R., Marquardt, H. & Todaro, G. J., Vaccinia virus encodes a polypeptide homologous to epidermal growth factor and transforming growth factor. *Nature,* **313** (1985) 491–2.

22. Bertholet, C., van Meir, E., ten Heggeler-Bordier, B. & Wittek, R., Vaccinia

virus produces late mRNAs by discontinuous synthesis. *Cell,* **50** (1987) 153–62.

23. Schwer, B., Visca, P., Vox, J. C. & Stunnenberg, H. G., Discontinuous transcription or RNA processing of vaccinia virus late mRNA results in a 5' poly A leader. *Cell,* **50** (1987) 163–9.

24. Wright, C. F. & Moss, B., *In vitro* synthesis of vaccinia virus late mRNA containing a 5' poly (A) leader sequence. *Proc. Nat. Acad. Sci., USA,* **84** (1987) 8883–7.

25. Schwer, B. & Stunnenberg, H. G., Vaccinia virus late transcripts generated *in vitro* have a poly (A) head. *EMBO J.,* **7** (1988) 1183–90.

26. Rosel, J. L., Earl, P. L., Weir, J. P. & Moss, B., Conserved TAAATG sequence at the transcriptional and translational start sites of vaccinia virus late genes deduced by structural and functional analysis of the HindIII H genome fragment. *J. Virol.,* **60** (1986) 436–49.

27. Weir, J. P. & Moss, B., Determination of the transcriptional regulatory region of a late vaccinia virus gene. *J. Virol.,* **61** (1987) 75–80.

28. Bertholet, C., Stocco, P., van Meir, E. & Wittek, R., Functional analysis of the 5' flanking sequence of a vaccinia virus late gene. *EMBO J.,* **5** (1986) 1951–7.

29. Moss, B., Rosenblum, E. N., Katz, E. & Grimley, P. M., Rifampicin: a specific inhibitor of vaccinia virus assembly. *Nature,* **224** (1969) 1280–4.

30. Mackett, M., Smith, G. L. & Moss, B., Vaccinia virus: a selectable eukaryotic cloning and expression vector. *Proc. Nat. Acad. Sci., USA,* **79** (1982) 7415–19.

31. Mackett, M., Smith, G. L. & Moss, B., General method for the production and selection of infectious vaccinia virus recombinants expressing foreign genes. *J. Virol.,* **49** (1984) 857–64.

32. Panicali, D. & Paoletti, E., Construction of poxviruses as cloning vectors: insertion of the thymidine kinase gene from herpes simplex virus into the DNA of infectious vaccinia virus. *Proc. Nat. Acad. Sci., USA,* **79** (1982) 4927–31.

33. Paoletti, E., Lipinskas, B. R., Samsonoff, C., Mercer, S. & Panicali, D., Construction of live vaccines using genetically engineered poxviruses: biological activity of vaccinia virus recombinants expressing the hepatitis B virus surface antigen and the herpes simplex virus glycoprotein D. *Proc. Nat. Acad. Sci., USA,* **81** (1984) 193–7.

34. Franke, C. A., Rice, C. M., Strauss, J. H. & Hruby, D. E., Neomycin resistance as a dominant selectable marker for selection and isolation of vaccinia virus recombinant. *Mol. Cell Biol.,* **5** (1985) 1918–24.

35. Boyle, D. B. & Coupar, B. E. H., A dominant selectable marker for the construction of recombinant poxviruses. *Gene,* **65** (1988) 123–8.

36. Falkner, F. G. & Moss, B., *Escherichia coli* gpt gene provides dominant selection for vaccinia virus open reading frame expression vectors. *J. Virol.,* **62** (1988) 1849–54.

37. Chakrabarti, S., Brechling, K. & Moss, B., Vaccinia virus expression vector: co-expression of beta-galactosidase provides visual selection of recombinant virus plaques. *Mol. Cell Biol.,* **5** (1985) 3403–9.

38. Rodriguez, J. F., Rodriguez, D., Rodriguez, J.-R., McGowan, E. B. & Esteban, M., Expression of the firefly luciferase gene in vaccinia virus: a highly sensitive gene marker to follow virus dissemination in tissues of infected animals. *Proc. Nat. Acad. Sci., USA,* **85** (1988) 1667–71.

39. Kieny, M. P., Lathe, R., Drillien, R., Sphener, D., Skory, S., Schmitt, D., Wiktor, T., Koprowski, H. & Lecocq, J. P., Expression of rabies virus glycoprotein from a recombinant vaccinia virus. *Nature*, **312** (1984) 163–6.

40. Fathi, Z., Sridhar, P., Facha, R. F. & Condit, R., Efficient targeted insertion of an unselected marker into the vaccinia virus genome. *Virology*, **155** (1986) 97–105.

41. Perkus, M. E., Panicali, D., Mercer, S. & Paoletti, E., Insertion and deletion mutants of vaccinia virus. *Virology*, **152** (1986) 285–97.

42. Flexner, C., Hugin, A. & Moss, B., Prevention of vaccinia virus infection in immunodeficient mice by vector-directed IL-2 expression. *Nature*, **330** (1987) 259–61.

43. Buller, R. M. L., Smith, G. L., Cremer, K., Notkins, A. L. & Moss, B., Decreased virulence of recombinant vaccinia virus expression vectors is associated with a thymidine kinase-negative phenotype. *Nature*, **317** (1985) 813–15.

44. Buller, R. M. L., Chakrabarti, S., Cooper, J. A., Twardzik, D. R. & Moss, B., Deletion of the vaccinia growth factor gene reduces virus virulence. *J. Virol.*, **62** (1988) 866–74.

45. Smith, G. L. & Moss, B., Infectious poxvirus vectors have capacity for at least 25,000 base pairs of foreign DNA. *Gene*, **25** (1983) 21–8.

46. Rosel, J. & Moss, B., Transcriptional and translational mapping and nucleotide sequence of a vaccinia virus gene encoding the precursor of the major core polypeptide 4b. *J. Virol.*, **56** (1985) 830–8.

47. Rice, C. M., Franke, C. A., Stauss, J. H. & Hruby, D. E., Expression of Sindbis virus structural proteins via recombinant vaccinia virus: synthesis, processing and incorporation into mature Sindbis virions. *J. Virol.*, **56** (1985) 227–39.

48. Chakrabarti, S., Robert-Guroff, M., Wong-Staal, F., Gallo, R. C. & Moss, B., Expression of HTLV-III envelope gene by recombinant vaccinia virus. *Nature*, **320** (1986) 535–7.

49. Smith, G. L., Mackett, M. & Moss, B., Infectious vaccinia virus recombinants that express hepatitis B virus surface antigen. *Nature*, **302** (1983) 490–5.

50. Wiktor, T. J., MacFarlan, R. I., Reagen, K. J., Dietzschold, B., Curtis, P. J., Wunner, W. H., Kieny, M.-P., Lathe, R., Lecocq, J.-P., Mackett, M., Moss, B. & Koprowswki, H., Protection from rabies by a vaccinia virus recombinant containing the rabies virus glycoprotein gene. *Proc. Nat. Acad. Sci., USA*, **81** (1984) 7194–8.

51. Sullivan, V. & Smith, G. L., Expression and characterization of herpes simplex virus type 1 (HSV-1) glycoprotein G (gG) by recombinant vaccinia virus: neutralization of HSV-1 infectivity with anti-gG antibody. *J. Gen. Virol.*, **68** (1987) 2587–98.

52. de la Salle, H., Altenburger, W., Elkaim, R., Dott, K., Dieterle, R., Cazenave, J. P., Tolstoshev, P. & Lecocq, J. P., Active α-carboxylated human factor VIII is expressed using recombinant DNA techniques. *Nature*, **316** (1985) 268–70.

53. Wittek, R., Hanggi, M. & Hiller, G., Mapping of a gene coding for a major late structural polypeptide on the vaccinia virus genome. *J. Virol.*, **49** (1984) 371–8.

54. Bennink, J. R., Yewdell, J. W., Smith, G. L. & Moss, B., Recognition of cloned influenza virus haemagglutinin gene products by cytotoxic T lymphocytes. *J. Virol.*, **57** (1986) 786–91.

55. Fuerst, T. R., Niles, E. G., Studier, F. W. & Moss, B., Eurkaryotic transient-expression system based upon recombinant vaccinia virus that synthesizes bacteriphage T7 RNA polymerase. *Proc. Nat. Acad. Sci., USA*, **83** (1986) 8122-6.

56. Fuerst, T. R., Earl, P. C. & Moss, B., Use of a hybrid vaccinia virus-T7 RNA polymerase system for expression of target genes. *Mol. Cell. Biol.*, **7** (1987) 2538-44.

57. Patel, D. D. & Pickup, D. J., Messenger RNAs of a strongly-expressed late gene of cowpox virus contain 5'-terminal poly (A) sequences. *EMBO J.*, **6** (1987) 3787-94.

58. Cranage, M. P., Kouzarides, T., Bankier, A. T., Satchwell, S., Weston, K., Tomlinson, P., Barrell, B., Hart, H., Bell, S. E., Minson, A. C. & Smith, G. L., Identification of the human cytomegalovirus glycoprotein B gene and induction of neutralising antibodies via its expression in recombinant vaccinia virus. *EMBO J.*, **5** (1986) 3057-63.

59. Cranage, M. P., Smith, G. L., Bell, S. E., Hart, H., Brown, C., Bankier, A. T., Tomlinson, P., Barrell, B. G. & Minson, A. C., Identification and expression of a human cytomegalovirus glycoprotein with homology to the Epstein–Barr virus BXLF2 product, varicella-zoster virus gpIII, and herpes simplex virus type 1 glycoprotein H. *J. Virol.*, **62** (1988) 1416-22.

60. Bennink, J. R., Yewdell, J. W., Smith, G. L., Moller, C. & Moss, B., Recombinant vaccinia virus primes and stimulates influenza virus haemagglutinin-specific cytotoxic T lymphocytes. *Nature*, **311** (1984) 578-9.

61. Bennink, J. R., Yewdell, J. W., Smith, G. L. & Moss, B., Anti-influenza virus cytotoxic T lymphocytes recognises the three viral polymerases and a nonstructural protein: responsiveness to individual viral antigens is major histocompatibility complex controlled. *J. Virol.*, **61** (1987) 1098-1102.

62. Yewdell, J. W., Bennink, J. R., Smith, G. L. & Moss, B., Influenza A virus nucleoprotein is a major target antigen for cross-reactive anti-influenza A virus cytotoxic T lymphocytes. *Proc. Nat. Acad. Sci., USA*, **82** (1985) 1785-9.

63. McMichael, A. J., Michie, C. A., Gotch, F. M., Smith, G. L. & Moss, B., Recognition of influenza virus nucleoprotein by human cytotoxic T lymphocytes. *J. Gen. Virol.*, **67** (1986) 719-26.

64. Gotch, F., McMichael, A., Smith, G. & Moss, B., Identification of viral molecules recognized by influenza-specific human cytotoxic T lymphocytes. *J. Exp. Med.*, **165** (1987) 408-16.

65. Bangham, C. R. M., Openshaw, P. J. M., Ball, L. A., King, A. M. Q., Wertz, G. M. & Askonas, B. A., Human and murine cytotoxic T cells to respiratory syncytial virus recognise the viral nucleocapsid (N), but not the major glycoprotein (G), expressed by vaccinia virus recombinants. *J. Immun.*, **137** (1986) 3973-7.

66. Coupar, B. E. H., Andrew, M. E., Both, G. W. & Boyle, D. B., Temporal regulation of influenza hemagglutinin expression in vaccinia. *Eur. J. Immun.*, **16** (1986) 1479-87.

67. Townsend, A., Bastin, J., Gould, K., Brownlee, G., Andrew, M., Boyle, D. B., Chan, Y. S. & Smith, G. L., Defective presentation to class I restricted CTL in vaccinia infected cells is overcome by enhanced degradation of antigen. *J. Exp. Med.*, **168** (1988) 1211-24.

68. Upton, C., Carrell, R. W. & McFadden, G., A novel member of the serpin superfamily is encoded on a circular plasmid-like DNA species isolated from rabbit cells. *FEBS Lett.*, **207** (1986) 115-20.

69. Pickup, D. J., Ink, B. S., Hu, W., Ray, C. A. & Joklik, W. K., Hemorrhage in lesions caused by cowpox virus is induced by a viral protein that is related to plasma protein inhibitors of serine proteases. *Proc. Nat. Acad. Sci., USA,* **83** (1986) 7698–702.

70. Tomley, F., Binns, M., Campbell, J. & Boursnell, M., Sequence analysis of an 11.2 kilobase, near-terminal BamHI fragment of fowlpox virus. *J. Gen. Virol.,* **69** (1988) 1025–40.

71. Boursnell, M. E. G., Foulds, I. J., Campbell, J. I. & Binns, M. M., Non-essential genes in the vaccinia virus HindIII K fragment: a gene related to serine protease inhibitors and a gene related to the 37K vaccinia virus major envelope antigen. *J. Gen. Virol.,* **69** (1988) 2995–3003.

72. Stephens, E. B., Compans, R. W., Earl, P. & Moss, B., Surface expression of viral glycoproteins is polarized in epithelial cells infected with vaccinia viral vectors. *EMBO J.,* **5** (1986) 237–45.

73. Smith, G. L., Levin, J., Palese, P. & Moss, B., Synthesis and cellular location of the ten influenza polypeptides individually expressed by recombinant vaccinia viruses. *Virology,* **160** (1987) 336–45.

74. Siomi, H., Shida, H., Nam, S. H., Nosaka, T., Maki, M. & Hatanaka, M., Sequence requirements for nucleolar localization of human T cell leukemia virus type 1 pX protein, which regulates viral RNA processing. *Cell,* **55** (1988) 197–209.

75. Moss, B., Smith, G. L., Gerin, J. L. & Purcell, R. H., Live recombinant vaccinia virus protects chimpanzees against hepatitis B. *Nature,* **311** (1984) 67–9.

76. Smith, G. L., Murphy, B. R. & Moss, B., Construction and characterisation of an infectious vaccinia virus recombinant that expresses the influenza virus haemagglutinin and induces resistance to influenza virus infection in hamsters. *Proc. Nat. Acad. Sci., USA,* **80** (1983) 7155–9.

77. Cremer, K., Mackett, M., Wohlenberg, C., Notkins, A. L. & Moss, B., Vaccinia virus recombinants expressing herpes simplex virus type 1 glycoprotein D prevents latent herpes in mice. *Science,* **228** (1985) 737–40.

78. Mackett, M., Yilma, T., Rose, J. & Moss, B., Vaccinia virus recombinants: expression of VSV genes and protective immunization of mice and cattle. *Science,* **227** (1985) 433–5.

79. Drillien, R., Sphener, D., Kirn, A., Giraudon, P., Buckland, R., Wild, F. & Lecocq, J.-P., Protection of mice from fatal measles encephalitis by vaccination with vaccinia virus recombinants encoding either the hemagglutinin or the fusion protein. *Proc. Nat. Acad. Sci., USA,* **85** (1988) 1252–6.

80. Spriggs, M. K., Collins, P. C., Tierney, E., London, W. T. & Murphy, F., Immunization with vaccinia virus recombinants that express the surface glycoproteins of human parainfluenza type 3 (PIV3) protects Patas monkeys against PIV3 infection. *J. Virol.,* **62** (1988) 1293–6.

81. Wertz, G. W., Stott, E. J., Young, K. K., Anderson, K. & Ball, L. A., Expression of the fusion protein of human respiratory syncytial virus from recombinant vaccina vectors and protection of vaccinated mice. *J. Virol.,* **61** (1987) 294–301.

82. Earl, P. L., Moss, B., Morris, R. P., Wehrl, Y. K., Nihsio, J. & Chesebro, B., T-lymphocyte priming and protection against Friend leukaemia by vaccinia-retrovirus env gene recombinant. *Science,* **234** (1986) 728–31.

83. Lathe, R., Kieny, M. P., Gerlinger, P., Clertant, P., Guizani, I., Cuzin, R. & Chambon, P., Tumour prevention and rejection with recombinant vaccinia virus. *Nature,* **326** (1987) 878–80.

84. Perkus, M. E., Piccini, A., Lipinskas, B. R. & Paoletti, E., Recombinant vaccinia virus: immunization against multiple pathogens. *Science,* **229** (1985) 981–4.

85. Lane, J. M., Ruben, F. L., Neff, J. M. & Millar, J. D., Complications of smallpox vaccination. 1968 Surveillance in the United States. *New Engl. J. Med.,* **281** (1969) 1201–8.

86. Kotwal, G. J. & Moss, B., Vaccinia virus encodes a secretory polypeptide structurally related to complement control proteins. *Nature,* **335** (1988) 176–8.

87. Gillard, S. D., Spehner, D., Drillien, R. & Kirn, A., Localization and sequence of a vaccinia virus gene required for multiplication in human cells. *Proc. Nat. Acad. Sci., USA,* **83** (1986) 5573–7.

88. Blancou, J., Kieny, M. P., Lathe, R., Lecocq, J. P., Pastoret, P. P., Soulebot, J. P. & Desmettre, P., Oral vaccination of the fox against rabies using a live recombinant vaccinia virus. *Nature,* **322** (1986) 373–5.

Chapter 8

THE EXPRESSION OF TISSUE-TYPE PLASMINOGEN ACTIVATOR AND RELATED ENZYMES

M. J. Browne, J. E. Carey, C. G. Chapman, I. Dodd, G. M. P. Lawrence & J. H. Robinson

Beecham Pharmaceuticals Research Division, Biosciences Research Centre, Great Burgh, Epsom, Surrey, UK

INTRODUCTION

Acute myocardial infarction (AMI) is a major cause of death in the Western World (Table 1). One of the main precipitating events in AMI is the formation of an occluding thrombus in a coronary artery. The thrombus prevents blood flow to the myocardium thus reducing oxygen supply which in turn leads to an, often fatal, infarction. It is now widely accepted that early removal of the blood clot, by dissolution of the fibrin network which holds it together, is of major benefit in reducing morbidity and mortality.[1] A number of fibrinolytic proteins are now in use in the clinic. One of these is tissue-type plasminogen activator (t-PA).[2] t-PA is a serine protease; however, it does not dissolve fibrin directly, instead it activates plasminogen, a plasma protein, by cleaving the arg_{561}–val_{562} peptide bond thus revealing the active enzyme, plasmin. Plasmin in turn dissolves fibrin and lyses the thrombus. The primary structure of t-PA was only established after cloning the mRNA as cDNA (Fig. 1).[3,4] The 527 amino acid protein, like many serine proteases, consists of two chains (A and B) linked by a protease-susceptible peptide bond. The B chain carries the catalytic centre, but somewhat unusually for a serine protease, t-PA appears to be active in both single and two-chain forms. The A chain is believed to be divided into four structural domains (the finger, the growth factor and two kringles). All the domains appear to have counterparts in many other proteins of the coagulation and fibrinolytic systems, consistent with the notion of generation of these complex proteins by the shuffling of exons.[5]

Initially t-PA was only available in minute quantities, *ex vivo* or from

117

Table 1
Major Causes of Death in the Western World

Cause of death	England/Wales (1984)[a]	W. Germany (1981)[b]	United States (1982)[b]
AMI	104 620	80 752	291 031
Malignant neoplasms	140 101	158 814	433 795
All causes	566 881	696 118	1 974 797

[a] Figures from Office of Population Census and Survey.
[b] Figures from World Health Organisation.

cell culture, and the main challenge was to obtain enough protein to evaluate it fully, both in the laboratory and in the clinic. More recently, as the attributes and deficiencies of t-PA began to emerge, the challenge arose to devise improved forms of t-PA and other fibrinolytic enzymes.

EXPRESSION OF NATIVE t-PA

Microbial Systems

Attracted by the potential for relative simplicity and economics of production many laboratories have examined expression in microbial systems. In our own case we chose to express t-PA as a fusion product using the pUC8 vector in *E. coli* (Fig. 2(a)). As had been the experience of

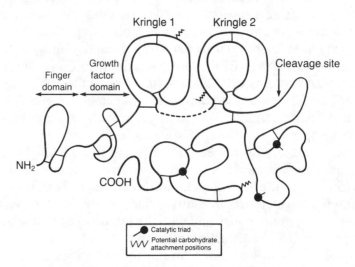

Fig. 1. Postulated structure of t-PA.

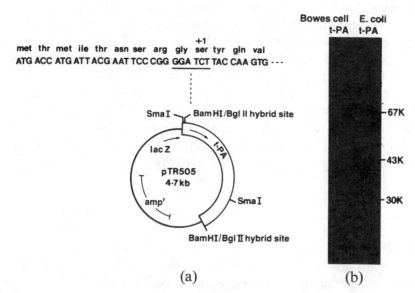

Fig. 2. Expression of t-PA in *E. coli*. (a) Expression construct, (b) zymographic analysis of native t-PA and *E. coli*-derived t-PA species.

several other groups,[3, 6] we found active protein could only be obtained by using a complex protocol comprising bacterial lysis followed by protein extraction/refolding in the presence of redox buffers. The yield achieved was a relatively poor 55 μg/litre of culture. Zymographic analysis showed the protein product to be (i) smaller than native t-PA (due to the absence of carbohydrate) and (ii) subject to a considerable degree of degradation (Fig. 2(b)). No group has reported efficient expression of t-PA from *E. coli*. It is thought that the processes of complex protein folding, disulphide bond formation and glycosylation cannot be accomplished efficiently by *E. coli*.

Other microbial expression systems, e.g. yeast[7] and filamentous fungi[8] have also been examined, but none has yet been widely adopted.

Mammalian Cell Culture

In the light of difficulties experienced using microbial systems most researchers have turned to higher eukaryotes for expression. Although some success has been achieved using transgenic animals[9] most groups now use one of a variety of mammalian cell culture systems for production of t-PA (see Chapter 6, this volume).

We have developed a flexible expression system based on the use of the vector pTR315.[10] The vector is designed to allow simple substitution of cDNA clones for subsequent expression. Without further modification the vector can be used for rapid, transient expression of t-PA (or other molecules) in a variety of mammalian cell lines. Using this system we can quickly produce small quantities of several different proteins for preliminary evaluation. The vector is also to allow direct incorporation of the 'expression cassette' into bovine papillomavirus vectors[10] or amplifiable vectors[11] for large-scale expression in stable cell lines. SDS-PAGE analysis of human t-PA preparations derived from a number of mammalian cell lines is shown in Fig. 3(a). Differences in mobility may be attributed to variations in glycosylation.

EXPRESSION OF t-PA MUTEINS AND HYBRID ENZYMES

Although t-PA appears to be a relatively attractive therapeutic agent there is no doubt that it should be possible to improve its properties.

Many laboratories are now using recombinant DNA techniques to dissect structure/function relationships within this complex molecule and to construct novel fibrinolytic agents, particularly t-PA muteins and hybrid enzymes. Such work is clearly highly dependent upon the ability of t-PA to undergo modification without detriment to activity. Results obtained in our laboratory underline the tolerance of t-PA to a wide variety of alterations.

Cleavage Site Mutations

The A/B-chain junction in t-PA is unusual amongst serine proteases in that (i) cleavage is apparently not a prerequisite for activity and (ii) a highly conserved hydrophobic amino acid residue found in all other serine proteases at the P'_2 site is substituted, in t-PA, by lysine. This region of the molecule is thus quite specific to t-PA and superficially might appear to be essential for activity. However, substitution of arg_{275} with gln or lys_{277} with ile or other modifications in this area[12] does not prevent expression of active protein (Fig. 3(b)).

Domain Deletion

In attempting to associate the properties of t-PA with the individual domains it is clearly desirable to be able to express molecules with various domains removed. We were, however, uncertain as to whether removal of one or more A-chain domains would prevent expression, for example by disrupting folding pathways, or perhaps result in an unstable

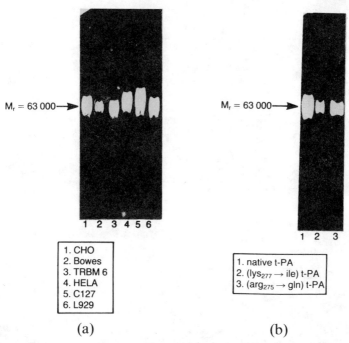

(a)   (b)

Fig. 3. Analysis of recombinant t-PA molecules using SDS-PAGE followed by fibrin zymography: (a) native t-PA derived from different mammalian cell lines, (b) mutant t-PA species expressed in Hela cells. (a) Recombinant t-PA was expressed using appropriate vectors in Hela cells (lane 4) or mouse C127 (lane 5) or L929 (lane 6) cells. The remaining tracks show t-PA harvested from Bowes melanoma cells (lane 2) and the recombinant Bowes Line TRBM6 (lane 3), and commercially available t-PA from Chinese hamster ovary cells (lane 1). (b) Plasmids encoding native t-PA (lane 1), $lys_{277} \rightarrow$ ile t-PA (lane 2) and $arg_{275} \rightarrow$ gln t-PA (lane 3) were constructed, expressed in Hela cells and analysed to show the fibrinolytically active species. The stained fibrin zymograms are shown. The apparent M_r of molecular weight markers are indicated.

molecule. In recent years we and others have used mutagenesis techniques to generate a number of deletion mutants,[10, 13, 14] for example (i) removing most of the A-chain, i.e. $---K_2B$ or (ii) simply excising a single internal domain, i.e. $F-K_1 K_2 B$ (Fig. 4). Both of these molecules are accurately synthesised and are active as judged by zymographic analysis (Fig. 5(a)).

It has been the experience of many groups working in this area that any or all A-chain domains can be removed without destroying activity, demonstrating that the B chain is catalytically autonomous.[15]

Domain Duplication

This technique allows us to probe the flexibility of the t-PA structure still further. We have created a large number of molecules with duplicated

122   *M. J. Browne, J. E. Carey, C. G. Chapman, I. Dodd, G. M. P. Lawrence and J. H. Robinson*

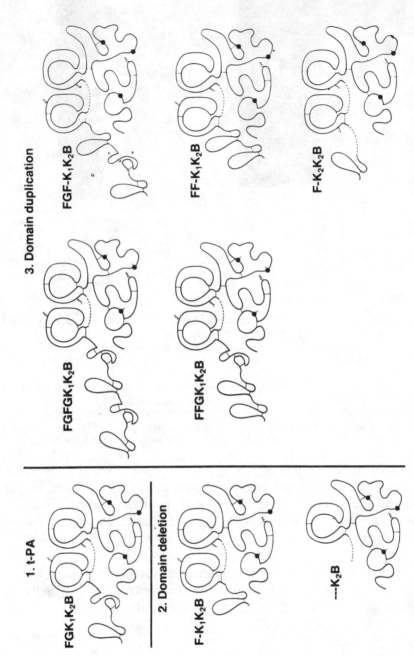

Fig. 4. Structures of novel t-PA species produced by domain deletion or domain duplication.

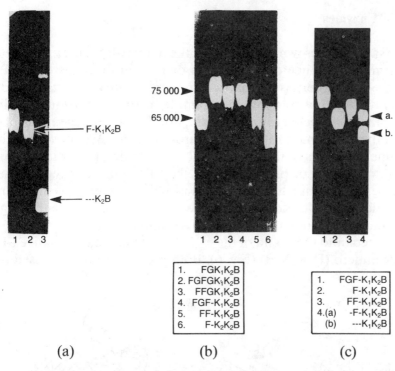

Fig. 5. SDS-PAGE followed by fibrin zymography of various t-PA mutants. (a) A chain domain deletions. Conditioned media from Hela cells transfected with plasmid encoding either $F-K_1K_2B$ or $---K_2B$ were purified on zinc chelate and lysine Sepharose. The major fibrinolytic species in each purified preparation is shown. Lane 1 contains a native t-PA marker. (b) Domain duplication muteins of t-PA. All species were synthesised in the Hela cell transient expression system and purified using zinc chelate and lysine Sepharose. The major fibrinolytic species in each purified preparation is shown. The two major species in lane 6 possibly represent different levels of glycosylation of $F-K_2K_2B$. The approximate apparent M_r of two molecular weight markers is indicated. (c) Degraded species present in a purified preparation of $FF-K_1K_2B$. Conditioned medium from Hela cell cultures transfected with a plasmid encoding $FF-K_1K_2B$ was fractionated by lysine Sepharose chromatography. Fibrinolytic markers of known structure are shown in lanes 1 ($FGF-K_1K_2B$) and 2 ($F-K_1K_2B$). Nascent $FF-K_1K_2B$ is shown in lane 3, degraded species are shown in lane 4. The presumed structures of the degraded species are $F-K_1K_2B$ and $--K_1K_2B$.

domains[16] exemplified in Fig. 4. Zymographic analysis again shows that expression is not compromised (Fig. 5(b)). However we have seen signs that all modifications may not be permissible, thus, on occasion, degradation products may be seen in the harvest medium of $FF-K_1K_2B$ (Fig. 5(c)). Presumably part(s) of this novel structure is (are) more accessible to proteolytic breakdown than the parent molecule.

Hybrid Enzymes

At the same time as work proceeds on the native t-PA molecule efforts are being made to splice other proteins or protein domains onto t-PA to improve its properties. In particular there is great interest in adding antibody targeting to t-PA, for example, using monoclonal antibody determinants directed towards fibrin,[17] or to combine elements of t-PA with other plasma proteins, e.g. urokinase.[18-21] We have sought to examine the limits of this approach by combining t-PA with the highly complex A chain of plasminogen. Like t-PA, plasminogen is a serine protease; the A chain, however, is composed almost entirely of five triple-disulphide bonded kringle structures (Fig. 6). cDNA encoding the plasminogen signal sequence and the A-chain region (amino acids 1–544) was spliced onto the 5' end of the cDNA encoding mature native t-PA or a t-PA mutein (F–$K_1 K_2 B$) (Fig. 6). Both proteins were expressed in the

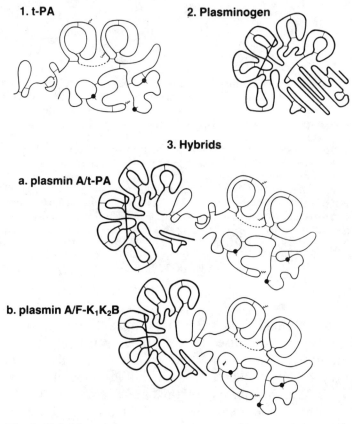

Fig. 6. Hybrid enzyme structures based on t-PA and plasminogen.

Hela cell transient system at a yield, based on activity, comparable with native t-PA. Zymographic analysis showed that full sized protein products could be obtained (Fig. 7).

This study underlines the fact that even apparently highly complex novel molecules can be expressed successfully in mammalian cell culture. A note of caution should be entered here since it is clear that not all hybrid proteins can be expressed efficiently. Thus in a recent study of t-PA/urokinase hybrid structures[19] it was found that while t-PA was present in harvest medium at a concentration of 16 ng/ml two t-PA/urokinase hybrids were only present at approximately 0·5 ng/ml.

This is a rapidly moving area and many other forms of recombinant hybrid enzymes[22] may be anticipated.

EXPRESSION OF GENOMIC DNA

The preceding sections of this chapter illustrate well the potential of cDNA-based mutagenesis programmes to produce novel protein structures. As well as using this 'classical' cDNA route we also used the t-PA gene itself.

The Bowes melanoma cell line constitutively secretes t-PA at a modest level. We isolated the whole 35kb t-PA gene on a single cosmid[23] and inserted extra copies into the Bowes cell, creating a new stable high-yielding line – TRBM6.[24] The high level of synthesis allowed us to isolate and characterise two novel t-PA species normally present in only trace amounts[25] (Fig. 8). Identification of these species indicates that post-transcriptional events can influence the primary sequence of the expressed t-PA protein.

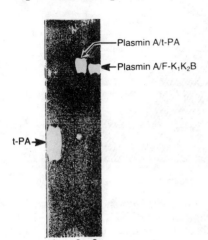

Plasmin A/t-PA

Plasmin A/F-K_1K_2B

t-PA

1 2 3

Fig. 7. Expression of hybrid plasminogen activators. SDS-PAGE followed by fibrin zymography of two complex hybrid plasminogen activators. Lane 1 is t-PA used as a marker. Lanes 2 and 3 are, respectively, plasmin A-chain/t-PA and plasmin A-chain/F-K_1K_2B; both show the major fibrinolytically active species present in conditioned medium from Hela cells transfected with plasmid coding for the respective hybrids.

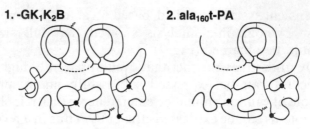

Fig. 8. Natural variants of t-PA.

$-GK_1K_2B$

We had identified the presence of a minor protein component in TRBM6-conditioned medium with $M_r = 56\,000$, compared to $M_r = 63/65\,000$ native t-PA. We had originally considered it likely that this represented a proteolytic breakdown product. Purification of this material (Fig. 9) allowed protein sequence analysis. This indicated that the molecule was probably produced from an mRNA species lacking the exon encoding the finger domain.[25]

ala_{160} t-PA

A further protein species of $M_r = 38\,000$ was also identified by zymography and this too was purified. Protein sequencing showed that the N-terminus corresponds to ala_{160}. The mechanism by which this molecule is produced is unclear, since (i) ala_{160} could not be generated

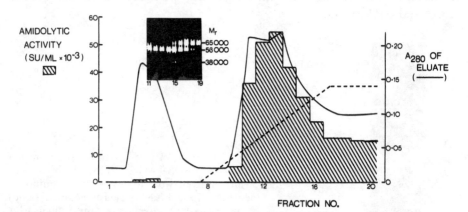

Fig. 9. Chromatography of apparent $M_r = 56\,000$ and $65\,000$ t-PA species on Heparin-Sepharose CL6B. t-PA was desorbed using a linear gradient of NH_4HCO_3 (0·05–1 M). Fractions were assayed using S2288 and by fibrin zymography (reproduced with permission).

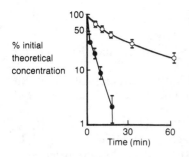

Fig. 10. Clearance of t-PA and a t-PA mutein *in vivo*. The plasma concentrations of t-PA or t-PA mutein were measured using a fibrin-plate (activity) assay.[10] ●, native t-PA; ○, F–$K_1 K_2 B$ t-PA mutein. Both proteins were expressed using bovine papillomavirus vectors in mouse C127 cells. Purification used zinc chelate and lysine Sepharose chromatography. Both proteins were reversibly acylated at the active site using AP-DAB[10] (reproduced with permission).

directly by an alternative mRNA splicing pathway and (ii) simple proteolysis is not sufficient as ala_{160} lies inside the disulphide cross-linked kringle 1 domain. It seems most likely that the final stage in the generation of ala_{160} t-PA must involve disulphide bond interchange as well as proteolysis. Such a mechanism has also been proposed to explain the production of microplasmin from plasmin.[26]

CONCLUSIONS AND PROSPECTS

By using cDNA and genomic DNA approaches we have been able to produce a wide range of novel t-PA structures for analysis. It is evident from the literature[13, 14, 27, 28] that many more molecules are under investigation in other laboratories throughout the world.

Most groups are working both to increase our understanding of the functions of t-PA and to produce improved variants. Surprisingly, even at this early stage, we can point to some success in terms of manipulating the properties of t-PA. Thus, one of the main drawbacks associated with t-PA is its very rapid loss from the circulation; the removal is mediated by a non-saturable hepatic clearance mechanism. We have found that simple excision of the growth factor domain (i.e. F–$K_1 K_2 B$) disrupts the clearance process and produces a molecule which is cleared considerably more slowly than t-PA *in vivo*[10] (Fig. 10). Doubtless many more improvements will be revealed in the near future.

ACKNOWLEDGEMENTS

The authors would like to thank C. Entwisle and B. Reavy for oligonucleotide synthesis, K. Carr for transient expression of the proteins and S. Cooke for assistance with protein purification.

REFERENCES

1. Marder, V. J. & Sherry, S., Thrombolytic therapy: current status. *New Engl. J. Med.*, **318** (1988) 1512–20, 1585–95.
2. Rijken, D. C. & Collen, D., Purification and characterisation of the plasminogen activator secreted by human melanoma cells in culture. *J. Biol. Chem.*, **256** (1981) 7035–41.
3. Pennica, D., Holmes, W. E., Kohr, W. J., Harkins, R. N., Vehar, G. A., Ward, C. A., Bennett, W. F., Yelverton, E., Seeburg, P. H., Heyneker, H. L., Goeddel, D. V. & Collen, D., Cloning and expression of human tissue-type plasminogen activator cDNA in *E. coli. Nature,* **301** (1983) 214–21.
4. Banyai, L., Varadi, A. & Patthy, L., Common evolutionary origin of the fibrin-binding structures of fibronectin and tissue-type plasminogen activator. *FEBS Lett.,* **163** (1983) 37–41.
5. Patthy, L., Evolution of the proteases of blood coagulation and fibrinolysis by assembly of molecules. *Cell,* **41** (1985) 657–63.
6. Harris, T. J. R., Patel, T., Marston, F. A. O., Little, S., Emtage, J. S., Opdenakker, G., Volckaert, G., Rombauts, W., Billiau, A. and DeSomer, P., Cloning of cDNA coding for human tissue-type plasminogen activator and its expression in *E. coli. Molec. Biol. Med.,* **3** (1986) 279–92.
7. Lemontt, J. F., Wei, C.-H. & Dackowski, W. R., Expression of active human uterine tissue plasminogen activator in yeast. *DNA,* **4** (1985) 419–28.
8. Upshall, A., Kumar, A. A., Bailey, M. C., Parker, M. D., Favreau, M. A., Lewison, K. P., Joseph, M. L., Maraganore, J. M. & McKnight, G. L., Secretion of active human tissue plasminogen activator from the filamentous fungus *Aspergillus nidulans. Bio/Technology,* **5** (1987) 1301–3.
9. Gordon, K., Lee, E., Vitale, J. A., Smith, A. E., Westphal, H. & Hennighausen, L., Production of human tissue plasminogen activator in transgenic mouse milk. *Bio/Technology,* **5** (1987) 1183–7.
10. Browne, M. J., Carey, J. E., Chapman, C. G., Tyrrell, A. W. R., Entwisle, C., Lawrence, G. M. P., Reavy, B., Dodd, I. & Robinson, J. H., A tissue-type plasminogen activator mutant with prolonged clearance *in vivo. J. Biol. Chem.,* **263** (1988) 1599–602.
11. Bebbington, C. R. & Hentschel, C. G., The use of vectors based on gene amplification for the expression of cloned genes in mammalian cells. In *DNA Cloning, a Practical Approach*, Vol. III, ed. D. M. Glover, IRL Press, Oxford, 1987, pp. 163–88.
12. Tate, K. M., Higgins, D. L., Holmes, W. E., Winkler, M. E., Heyneker, H. L. & Vehar, G. A., Functional rôle of proteolytic cleavage at arginine$_{275}$ of human tissue plasminogen activator as assessed by site-directed mutagenesis. *Biochemistry,* **26** (1987) 338–43.
13. Harris, T. J. R., Second generation plasminogen activators. *Prot. Eng.,* **1** (1988) 449–58.
14. Krause, J., Catabolism of tissue-type plasminogen activator, its variants, mutants and hybrids. *Fibrinolysis,* **2** (1988) 133–42.
15. Dodd, I., Fears, R. & Robinson, J. H., Isolation and preliminary characterisation of active B-chain of recombinant tissue-type plasminogen activator. *Thromb. Haem.,* **55** (1988) 133–42.

16. Browne, M. J. & Robinson, J. H., European Patent Application Publication, 1987, 0 241 210.
17. Schnee, J. M., Runge, M. S., Matsueda, G. R., Hudson, N. W., Seidman, J. G., Haber, E. & Quertermous, T., Construction and expression of a recombinant antibody-targeted plasminogen activator. *Proc. Nat. Acad. Sci.,* **84** (1987) 6904–8.
18. Lee, S. G., Kalyan, N., Wilhelm, J., Hum, W.-T., Rapparport, R., Cheng, S.-M., Dheer, S., Urbano, C., Hartzell, R. W., Ronchetti-Blume, M., Levner, M. & Hung, P. P., Construction and expression of hybrid plasminogen activators prepared from tissue-type plasminogen activator and urokinase-type plasminogen activator genes. *J. Biol. Chem.,* **263** (1988) 2917–24.
19. de Vries, C., Veerman, H., Blasi, F. & Pannekoek, H., Artificial exon shuffling between tissue-type plasminogen activator and urokinase. *Biochemistry,* **27** (1988) 2565–72.
20. Pierard, L., Jacobs, P., Gheysen, D., Hoylaerts, M., Andre, B., Topisirovic, L., Cravador, A., de Foresta, F., Herzog, A., Collen, D., de Wilde, M. & Bollen, A., Mutant and chimeric recombinant plasminogen activators. *J. Biol. Chem.,* **262** (1987) 11771–8.
21. Gheysen, D., Lijnan, H. R., Pierard, L., de Foresta, F., Demarsin, E., Jacobs, P., de Wilde, M., Bollen, A. & Collen, D., Characterisation of a recombinant fusion protein of the finger domain of tissue-type plasminogen activator with a truncated single chain urokinase-type plasminogen activator. *J. Biol. Chem.,* **262** (1987) 11779–84.
22. Robinson, J. H., Dodd, I., Esmail, A., Ferres, H. & Nunn, B., Slow clearance of acylated, hybrid thrombolytic enzymes. *Thromb. Haem.,* **59** (1988) 421–5.
23. Browne, M. J., Tyrrell, A. W. R., Chapman, C. G., Carey, J. E., Glover, D. M., Grosveld, F. G., Dodd, I. & Robinson, J. H., Isolation of a human tissue-type plasminogen activator genomic DNA clone and its expression in mouse L cells. *Gene,* **33** (1985) 279–84.
24. Browne, M. J., Dodd, I., Carey, J. E., Chapman, C. G. & Robinson, J. H., Increased yield of human tissue-type plasminogen activator obtained by means of recombinant DNA technology. *Thromb. Haem.,* **54** (1985) 422–4.
25. Dodd, I., Nunn, B. & Robinson, J. H., Isolation, identification and pharmacokinetic properties of human tissue-type plasminogen activator species. *Thomb. Haem.,* **59** (1988) 523–8.
26. Wu, H.-L., Shi, G.-Y., Wohl, R. C. & Bender, M. L., Structure and formation of microplasmin. *Proc. Nat. Acad. Sci.,* **84** (1987) 8793–5.
27. Pannekoek, H., de Vries, C. & van Zonneveld, A.-J., Mutants of human tissue-type plasminogen activator: structural aspects and functional properties. *Fibrinolysis,* **2** (1988) 123–32.
28. Gething, M. J., Adler, B., Boose, J.-A., Gerard, R. D., Madison, E. L., McGookey, D., Meidell, R. S., Roman, L. M. & Sambrook, J., Variants of human tissue-type plasminogen activator that lack specific structural domains of the heavy chain. *EMBO J.,* **7** (1988) 2731–40.

Chapter 9

EXPRESSION OF RECOMBINANT FACTOR VIII MOLECULES IN MAMMALIAN CELLS

PIERRE MEULIEN, THÉRÈSE FAURE & ANDRÉA PAVIRANI

Transgène S.A., Strasbourg, France

INTRODUCTION

Factor VIII (FVIII) is an essential cofactor involved in blood clotting. It functions in the intrinsic pathway of coagulation in the step where the activated form of factor IX (FIXa) activates in its turn factor X to FXa. Lack or malfunction of FVIII results in the chromosome X-linked bleeding disorder haemophilia A. In blood, FVIII circulates as a complex with von Willebrand factor (VWF) which serves as carrier molecule stabilising FVIII procoagulant activity and may target FVIII to damaged sites. At present, haemophiliacs are treated by replacement therapy with plasma derived products which pose well documented problems associated with low purity and risk of contamination by viral agents such as those causing hepatitis A, B, non-A non-B and AIDS.

For these reasons, FVIII was an obvious target for recombinant DNA technology. The FVIII gene and cDNA were cloned originally by three independent groups[1-4] and many laboratories, including our own, are now producing recombinant FVIII (rFVIII) from mammalian cell culture. Due to the complexity and size of the FVIII protein, the primary translation product of which is 2351 amino acids (aa), mammalian cells are most certainly the only suitable host for its expression. A wealth of information concerning the structure/function relationships of FVIII was obtained due to the elegant biochemistry and molecular biology which culminated in its cloning thus enabling the elucidation of the FVIII amino acid sequence. Internal homologies revealed a domain structure with the pattern: A_1-A_2-B-A_3-C_1-C_2 (Refs 2 and 5). The unique

B domain lies between the heavy and light chains (see Fig. 1) and is heavily glycosylated.

The FVIII protein has been shown to be poorly secreted from heterologous cells and Dorner *et al.*[6] have recently shown that, in CHO (Chinese hamster ovary) cells at least, this is due to FVIII binding to a glucose regulated protein called BIP, the latter blocking FVIII in the endoplasmic reticulum. This, along with the fact that the B domain does not partake in the FVIII active complex (Fig. 1) has prompted several groups to make genetically engineered variants of FVIII. Two approaches have been used, firstly, separately expressing the two functional domains[7,8] and secondly producing variants deleted to a greater or lesser extent in the B domain.[9-13] In this chapter, we describe briefly the expression of complete FVIII in heterologous cells and go on to describe more recent data on some variant molecules which may well be candidates for second generation therapeutics for haemophilia A.

EXPRESSION OF COMPLETE rFVIII

In the first instance we used a vaccinia virus expression system to study the expression of rFVIII in various cell types.[14] It was clear from this study

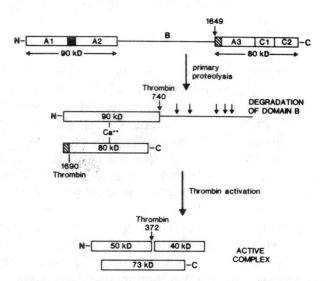

Fig. 1. Proteolytic processing of FVIII after Toole *et al.*[2] and Vehar *et al.*[5] The primary cleavage by an unknown protease occurs between aa 1648 and 1649. Degradation of the B domain is then thought to occur prior to activation by thrombin (IIa). The association between heavy and light chains is mediated by a metal cation, probably Ca^{2+}. Reproduced with permission from Meulien *et al. Protein Engineering,* **2**(4) (1988) 301–6.

that (i) the ability or inability to express biologically active rFVIII was cell type dependent and (ii) the addition of vWF preparations to the cell supernatant preserves FVIII activity over a period of 48 h thus stabilising this fragile molecule. We have since established CHO derived cell lines expressing rFVIII at high levels allowing purification and characterisation of the recombinant protein.

In order to establish permanent cell lines producing rFVIII we used vectors which enable either the immediate chromosomal integration of an elevated copy number of the expression unit containing the FVIII sequence, or vectors allowing low copy number integration followed by gene amplification. In the first case the vector contains a dicistronic unit comprising the FVIII cDNA followed by the XGPRT (xanthine guanine phospho ribosyl transferase) selection gene (pTG 1020 based on pTG381; see Refs 13 and 15). A mouse mitochondrial DNA sequence responsible for high copy number integration in the host chromosome is present in the construction.[16] In the second case (pTG1566), the dicistronic unit includes the DHFR (dihydrofolate reductase) gene to enable amplification using increasing concentration of methotrexate (MTX) (no mouse mitochondrial DNA sequences are present in this construction). CHO or CHO DHFR⁻ cells were transfected with pTG1020 and pTG1566 respectively, and after selection several clones (pTG1020) or mixtures of clones (pTG1566) were isolated and characterised for the presence of active rFVIII in their supernatants. FVIII was indeed detected in both cases.

Supernatants from pTG1020 derived clones were used for preliminary characterisation of the FVIII molecule while mixtures of clones derived from the transfection with the pTG1566 were subjected to increasing concentrations of MTX resulting in the progressive amplification of the copy number of exogenous DNA (as revealed by Southern blot analysis; see Fig. 2) and the corresponding progressive amplification of the FVIII expression of at least 10 fold. Several high producing clones were chosen and characterised for long term stability of growth and FVIII production in the absence of any selective agent in the medium. Under these conditions one particular population of cells was isolated and grown as a batch culture. This allowed us to harvest large quantities of rFVIII containing supernatant. rFVIII was immunopurified and its biochemical and functional properties were compared with those of plasma derived (pd) FVIII. In particular the SDS-PAGE profile, the immunorecognition by antibodies directed against human FVIII, the thrombin activation profile and the NH_2 terminal aa sequence revealed no differences between the two molecules (data not shown).

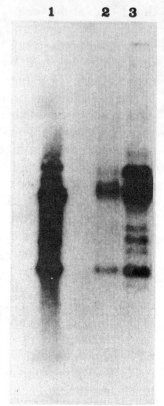

Fig. 2. Southern blots of total cellular EcoRI digested DNA from cells derived from the pTG1566 transfection. Lane 1: plasmid pTG1566; Lane 2: DNA from mixture No. 7 selected on nucleoside free medium; Lane 3: DNA from a clone isolated after amplification (using 30 nM methotrexate) of mixture No. 7. This FVIII gene amplification corresponded to a 10-fold increase in FVIII expression.

EXPRESSION OF FVIII VARIANTS

If we look at the processing of FVIII during its activation (by thrombin) and inactivation (by thrombin, FXa and activated protein C) it becomes apparent that the large B domain (Fig. 1) consisting of nearly 1000 amino acids does not play a role in the active complex. Many groups have used this feature of FVIII to derive new molecules which would be easier to produce and purify than the complete FVIII molecule. We attempted to express the two functional domains of FVIII namely the heavy (90) kD and light (80 kD) chains on separate vectors in the same cell to see if procoagulant activity could be demonstrated. Indeed FVIII activity can be observed in this system[8] however owing to the inefficiency of heavy and light chain association after expression, an essential requirement for FVIII function, much lower activity levels than expected were observed.

Another approach has been to delete portions of the B domain and express truncated molecules in mammalian cells.[9-13] We have made several such FVIII variants and expressed them both in BHK (baby

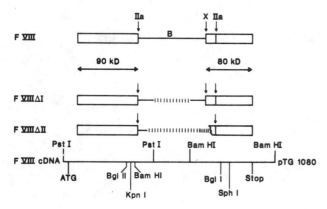

Fig. 3. Representation of FVIII and two deletion derivatives, FVIIIΔI in which DNA encoding amino acids 868–1582 is missing and FVIIIΔII which lacks the DNA encoding amino acids 771–1666. Deletions are denoted by (॥॥॥॥). Cleavage sites for thrombin (IIa) and that of the unknown protease (X) which cleaves between amino acid 1648 and 1649 (the latter deleted in FVIIIΔII) are shown. At the bottom the FVIII cDNA in the plasmid pTG1080[14] is represented showing several restriction enzyme sites. The FVIIIΔII construction was made by loop out mutagenesis using a 36-mer synthetic oligonucleotide.
   Reproduced with permission from Meulien *et al. Protein Engineering,* **2**(4) (1988) 301–6.

hamster kidney) cells using the vaccinia virus system and in permanent CHO derived cell lines. Two such molecules are shown in Fig. 3. The first deletion FVIIIΔI in which aa 868–1562 are missing retains the primary processing site at aa 1648/1649 the cleavage of which is suspected to occur inside the cell. The second molecule (FVIIIΔII) abolishes this site, the deletion extending from aa 771–1666. Both molecules exhibit increased expression levels compared to complete FVIII. In order to characterise further FVIIIΔII, permanent CHO derived cell lines were established.

As expected, the destruction of the cleavage site at aa 1648/1649 results in an uncleaved single chain FVIII molecule being secreted from the cell. Figure 4 shows an immunoprecipitation (using an anti-FVIII monoclonal antibody) of conditioned medium after metabolic labelling of cells. Whereas complete FVIII (lane 2) is visualised as several bands of equal intensity varying from ~200 to 80 kD, FVIIIΔII migrates for the most part, as a single band of the expected size 160–170 kD. This molecule could be therefore expected to exhibit increased stability in cell culture and during the purification process.

Deletions into the FVIII B domain which approach the thrombin cleavage site at aa 1689/1690 may be expected to impair or interfere with the thrombin activation of the molecule. We therefore compared thrombin activation profiles of complete FVIII and FVIIIΔII. In Fig. 5 it can be seen that whereas plasma derived or rFVIII are activated in a similar manner, FVIIIΔII is activated to a three-fold greater extent. This

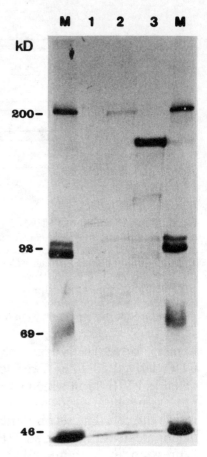

Fig. 4. Immunoprecipitation of rFVIII (lane 2) and FVIII △ II (lane 3) from conditioned medium of metabolically labelled CHO cells expressing the appropriate molecule. Lane 1 shows the control immunoprecipitation using conditioned medium from untransfected CHO cells. M = molecular weight markers. Reproduced with permission from Meulien *et al. Protein Engineering,* 2(4) (1988) 301–6.

result, we feel, may indicate an improved biological property of this molecule.

Foster *et al.*[17] have mapped a major vWF binding site to part of the acidic region aa 1649–1689. We have demonstrated that FVIII △ II retains the ability to bind vWF in an in-vitro assay[13] and so this molecule could be expected to complex with vWF *in vivo* and therefore exhibit a normal half life in plasma. FVIII △ II is presently under test in a haemophilic dog model.

CONCLUSIONS

The cloning and expression of the FVIII cDNA have opened the door to the development of safer treatment for haemophilia A patients. Already rFVIII has been used successfully in limited clinical trials in USA[18] which are now being extended to Europe. The advances in the

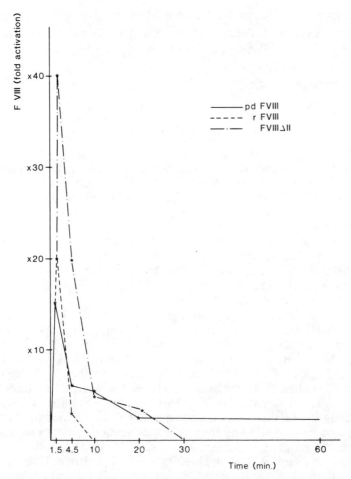

Fig. 5. Thrombin activation profile (average of 10 determinations) of pd FVIII, rFVIII and FVIII△II showing increased activation in the case of FVIII△II compared to plasma derived or recombinant molecules. Reproduced with permission from Meulien *et al.* *Protein Engineering,* **2**(4) (1988) 301-6.

understanding of FVIII function has also allowed the development of new molecules which may give rise to second generation therapeutics. However stringent trials, especially regarding the potential immunogenicity of these molecules, will be required due to the fact that FVIII replacement therapy consists of repeated administration over a long time period.

ACKNOWLEDGEMENTS

We would like to acknowledge all our colleagues at Transgène and at the blood transfusion centres of Strasbourg, Lille and Paris who participated in this work and N. Poujol for secretarial assistance. We especially thank Jean-Pierre Lecocq for encouragement and Gérard Jacquin for co-ordinating the collaboration between Transgène and the Centres National de Transfusion Sanguine within which this work was made possible through funding from the CNTS.

REFERENCES

1. Wood, W. I., Capon, D. J., Simonsen, C. C., Eaton, D. L., Gistchier, J., Keyt, B., Seeburg, P. H., Smith, D. H., Hollingshead, P., Wion, K. L., Delwart, E., Tuddenham, E. G. D., Vehar, G. A. & Lawn, R. M., Expression of active human factor VIII from recombinant DNA clones. *Nature*, **312** (1984) 330–7.
2. Toole, J. J., Knopf, J. L., Wosney, J. M., Sultzman, L. A., Bucker, J. L., Pittman, D. D., Kaufman, R. J., Brown, E., Shoemaker, C., Orr, E. C., Amphlett, G. W., Foster, B. W., Coe, M. L., Knuston, G. J., Fass, D. N. & Hewick, R. M., Molecular cloning of a cDNA encoding human antihaemophilic factor. *Nature*, **312** (1984) 342–7.
3. Gistchier, J., Wood, W. I., Goralka, T. M., Wion, K. L., Chen, E. Y., Eaton, D. H., Vehar, G. A., Capon, D. J. & Lawn, R. M., Characterization of the human factor VIII gene. *Nature*, **312** (1984) 326–30.
4. Truett, M. A., Blacher, R., Burke, R. L., Caput, D., Chu, C., Dina, D., Hartog, K., Kuo, C. H., Masiarz, F. R., Merryweather, J. P., Najarian, R., Pachl, C., Potter, S. J., Puma, J., Quiroga, M., Rall, L. B., Randolph, A., Urdea, M. S., Valenzuela, P., Dahl, H. H., Favalaro, J., Hansen, J., Nordfang, O. & Ezban, M., Characterization of the polypeptide composition of human factor-VIIIc and the nucleotide sequence and expression of the human kidney complementary DNA. *DNA*, **4** (1985) 333–49.
5. Vehar, G. A., Keyt, B., Eaton, D., Rodriguez, H., O'Brien, D. P., Rotblat, F., Oppermann, H., Keck, R., Wood, W. I., Harkins, R. N., Tuddenham, E. G. D., Lawn, R. M. & Capon, D. J., Structure of human factor VIII. *Nature*, **312** (1984) 337–42.
6. Dorner, A. J., Bole, D. G. & Kaufman, R. J., The relationship of N-linked glycosylation and heavy chain-binding protein association with the secretion of glycoproteins. *J. Cell Biol.*, **105** (1987) 2665–74.
7. Burke, R. L., Pachl, C., Quiroga, M., Rosenberg, S., Haigwood, N., Nordfang, O. & Ezban, M., The functional domains of coagulation factor VIII:c. *J. Biol. Chem.*, **261** (1986) 12574–8.
8. Pavirani, A., Meulien, P., Harrer, H., Schamber, F., Dott, K., Villeval, D., Cordier, Y., Wiesel, M.-L., Mazurier, C., Van de Pol, H., Piquet, Y., Cazenave, J.-P. & Lecocq, J.-P., Two independent domains of factor VIII co-expressed using recombinant vaccinia viruses have procoagulant activity. *Biochem. Biophys. Res. Commun.*, **145** (1987) 234–40.
9. Toole, J. J., Pittman, D. D., Orr, E. C., Murtha, P., Wasley, L. C. & Kaufman, R. J., A large region ($\approx$95 kDa) of human factor VIII is dispensable for in vitro procoagulant activity. *Proc. Nat. Acad. Sci., USA*, **83** (1986) 5939–42.

10. Eaton, D. L., Wood, W. I., Eaton, D., Haas, P. E., Hollingshead, P., Wion, K., Mather, J., Lawn, R. M., Vehar, G. A. & Gorman, C., Construction and characterization of an active factor VIII variant lacking the central one-third of the molecule. *Biochemistry,* **25** (1986) 8343–7.
11. Sarver, N., Ricca, G. A., Link, J., Nathan, M. H., Newman, J. & Drohan, W. N., Stable expression of recombinant factor VIII molecules using a bovine papillomavirus vector. *DNA,* **6** (1987) 553–63.
12. Pittman, D. & Kaufman, R. J., Proteolytic requirements for thrombin activation of anti-hemophilic factor (factor VIII). *Proc. Nat. Acad. Sci., USA,* **85** (1988) 2429–33.
13. Meulien, P., Faure, T., Mischler, F., Harrer, H., Ulrich, P., Bouderbala, B., Dott, K., Sainte-Marie, M., Mazurier, C., Wiesel, M.-L., Van de Pol, H., Cazenave, J.-P., Courtney, M. & Pavirani, A., A new recombinant procoagulant protein derived from the cDNA encoding human factor VIII. *Prot. Engng,* **2** (1988) 301–6.
14. Pavirani, A., Meulien, P., Harrer, H., Schamber, F., Dott, K., Villeval, D., Cordier, Y., Wiesel, M.-L., Mazurier, C., Van de Pol, H., Piquet, Y., Cazenave, J.-P. & Lecocq, J.-P., Choosing a host cell for active recombinant factor VIII production using vaccinia virus. *Bio/Technology,* **5** (1987) 389–92.
15. Balland, A., Faure, T., Carvallo, D., Cordier, P., Ulrich, P., Fournet, D., De la Salle, H. & Lecocq, J.-P., Characterisation of two differently processed forms of human recombinant factor IX synthesised in CHO cells transformed with a polycistronic vector. *Eur. J. Biochem.,* **172** (1986) 565–72.
16. Lutfalla, G., Blanc, H. & Bertolotti, R., Shuttling of integrated vectors from mammalian cells to *Escherichia coli* is mediated by head-to-tail multimeric inserts. *Somat. Cell Mol. Genet.,* **11** (1985) 223–38.
17. Foster, P. A., Fulcher, L. A., Houghton, R. A. & Zimmerman, T., An immunogenic region within residues val[1670]–glu[1684] of the factor VIII light chain induces antibodies which inhibit binding of factor VIII to von Willebrand factor. *J. Biol. Chem.,* **263** (1988) 5230–4.
18. White II, G. C., McMillan, C. W., Kingdom, H. S. & Shoemaker, C. B., Use of recombinant antihemophilic factor in the treatment of two patients with classic hemophilia. *New Engl. J. Med.,* **320** (1989) 166–70.

Chapter 10

THE β-GLOBIN DOMINANT CONTROL REGION

FRANK GROSVELD, MIKE ANTONIOU, GREET BLOM VAN ASSENDELFT,
PHIL COLLIS, NIALL DILLON, DAVID R. GREAVES, OLIVIA HANSCOMBE,
JACKY HURST, MICHAEL LINDENBAUM, DALE TALBOT & MIGUEL VIDAL

Laboratory of Gene Structure and Expression, National Institute for Medical Research, Mill Hill, London, UK

The strongest evidence for the existence of an important control in the flanking region of the globin gene domain was provided by the analysis of human $\gamma\beta$-thalassaemias.[1,2] Patients with heterozygous Dutch $\gamma\beta$-thalassaemia have a deletion that removes 100 kb of DNA, leaving the β-globin gene and the promoter and enhancer regions intact. However, it abolishes expression of the deleted chromosome and leaves the gene in an inactive chromatin configuration.[3-5] The wild-type allele on the other chromosome is expressed at normal levels, indicating that there is no shortage of *trans*-acting factors. This suggests a *cis* effect on β-globin gene transcription, which could be caused by a loss of positive acting elements or by the juxtaposition of the intact β-globin gene and sequences that remain in an inactive chromatin configuration in erythroid cells. The first indication that positive acting sites may be involved in activation of the β-globin domain came with the observation of erythroid specific DNaseI hypersensitive sites that map 6–18 kb upstream from the ε-globin gene (Fig. 1).[6-8]

When this region is added to a human β-globin gene construct, it results in very high levels of human β-globin gene expression in transgenic mice, which is related to the copy number and independent of the integration site of the transgene (Fig. 1).[8] The level of human β-globin gene expression observed in foetal liver RNA of 13·5 day old embryos is far higher than that observed previously in experiments with transgenic mice.[9-11] Furthermore, the level of human β-globin gene expression per gene copy is very nearly equal to that of the mouse endogenous β^{maj} globin gene. The authors are, therefore, confident that the β-globin minilocus construct contains all the positive acting sequences required

Frank Grosveld, et al.

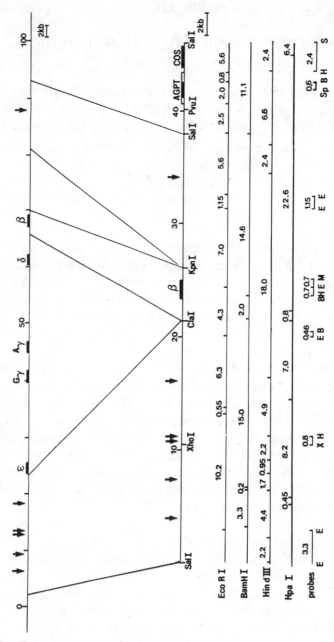

Fig. 1. The human β-globin gene locus. The human β-globin domain with all the functional genes is illustrated at the top. The β-globin minilocus leading to full expression of the β-globin gene in transgenic mice and MEL cells[8,12] is shown at the bottom; arrows indicate DNaseI super hypersensitive sites; sizes are in kilobases.

for activation of the human β-globin gene during development. When the same β-globin minilocus is transfected directly into erythroid cells (MEL and K562), high level β-globin expression is obtained.[12] The expression level per human β-globin gene is at a similar level to that of the endogenous mouse globin genes in the MEL cells. This effect is not obtained in non-erythroid L-cells, where low levels of expression are obtained similar to those obtained without flanking regions. This implies that the β-globin locus does not need to undergo a complete differentiation programme to be expressed at high levels. The presence of the flanking region also results in a dramatic stimulation of transcription of the non-erythroid specific promoter of the HSV thymidine kinase (tk) gene after differentiation of the MEL cells. This dramatic stimulation of the TK neo[r] gene on erythroid differentiation is not dependent on the presence of the β-globin gene promoter or enhancer sequences, as the β-globin gene can be replaced in the minilocus by a murine-human Thy-1 gene fragment and a similar stimulation of TK neo transcription is seen (Fig. 2 and see below). A similar stimulation of an SV40-neo gene is seen when it is integrated into the human β-globin locus by homologous recombination.[13] Recloning of the upstream hypersensitive sites on small fragments (6 kb total size) shows that both the β-globin and tk-neo gene are still expressed highly (Fig. 3),[14] in the case of the β-globin gene, at levels even higher than in the minilocus[12] and this may be caused by a

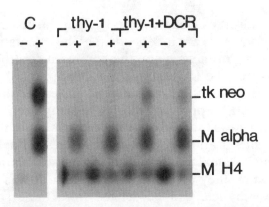

Fig. 2. Expression analysis of the human β-globin and Thy-1 MEL clones and populations by Northern blotting. A mouse–human hybrid Thy-1 gene was cloned between the ClaI and KpnI sites of the β-globin minilocus cosmid (lanes Thy-1 + DCR) or the cosmid vector pTCF (lanes Thy-1) and G418 resistant MEL cell populations were generated. 15 μg of RNA from uninduced (−) as well as induced (+) clones and populations was Northern blotted after gel electrophoresis and hybridised with a mixture of the TK neo probe, the mouse α-globin BamHI second exon probe (M alpha) and the mouse histone H4 probe (MH$_4$). RNA from untransfected MEL-C88 cells transfected with the β-globin minilocus was used as a positive control (C).

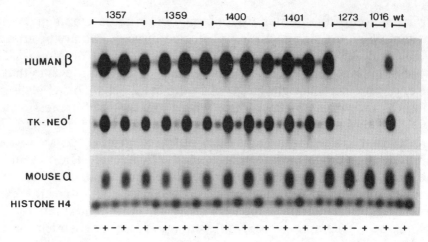

Fig. 3. Northern blot analysis of MEL cell populations stably transfected with the human β-globin microlocus.[14] The microlocus constructs contained the four 5′ hypersensitive sites linked to the 4·8 kb BglII fragment of the β-globin gene (2·0 kb, 1·9 kb, 1·5 kb and 1 kb, respectively, and the β-globin gene). 1359 contains the HSS, β-globin gene and TK-neo gene in a sense orientation from 5′ to 3′ in the construct. 1400 has the HSS inverted, 1401 has the β-globin gene inverted and 1357 has both the HSS and β-globin gene inverted (1273 does not contain the HSS region). After electrophoresis of 10 μg of RNA and blotting to nitrocellulose, the filters were probed with either a human β-globin probe or a tk-neo resistance gene probe. The filter initially probed with human β-globin was reprobed with both the mouse histone H4 probe and the mouse α-globin probe. Lanes marked 1016 contain RNA from a MEL cell clone stably transfected with the human β-globin 'minilocus'. Wt contains RNA from untransfected MEL C88 cells.

distance effect or by the deletion of the 3′ hypersensitive region. The very high transcription levels are independent of the orientation and relative positions of the β-globin gene and the dominant control region (DCR) and, in this respect, the DCR shows the same properties as a classical enhancer.[14] Deletion of the β-globin gene leaves the tk-neo gene fully active and deletion mapping of the DCR using both MEL cells and transgenic mice shows that the most important sites in the DCR are the hypersensitive sites 2 and 3 (Collis *et al.*, in press). Each of these have an approximately equal contribution to the stimulation of β-globin transcription when they are tested in isolation. It was found that the β-globin gene enhancer can be deleted from these small constructs without any adverse effect on the transcription levels of the β-globin gene (Antoniou *et al.*, unpublished). The DCR has also been shown to act on the embryonic and foetal globin genes in embryonic stem cells, MEL cells, K562 cells and transgenic mice (Lindenbaum, Dillon and Hurst, unpublished). Moreover, it can activate the α-globin gene and other non-erythroid genes such as Thy-1 in MEL cells and transgenic mice.[12]

Interestingly, all of the DNaseI hypersensitive sites have already formed before the induction of MEL cell differentiation and the start of globin transcription.[12] In L-cell populations stably transfected with the β-globin minilocus construct, the only DNaseI hypersensitive site reformed is site 3. The significance of this is presently unclear, but it suggests that the formation of site 3 may be dependent upon it being present in 'active' chromatin. This is assured in transfection experiments by virtue of G418 selection. Together with the data described above, this suggests that the DCR contains elements of more than one function whose action is required or affected at different times during the differentiation of erythroid cells. The DCR chromatin structure has already changed before the induction of differentiation, while a second inducible function is linked to the transcription of the genes.

The control of globin gene expression by the hypersensitive region is completely dominant and the sites are present at all stages of erythroid development.[6-8] There is a good possibility that this region controls the accessibility of the β-globin locus to *trans*-acting factors (see γβ-thalassaemia[3]), perhaps like the border regions flanking the DNaseI-sensitive domain originally described for the ovalbumin genes and chicken globin genes.[15, 16] It was originally speculated that these borders may contain nuclear matrix binding sites[17] and Harman and Higgs[18] have recently shown the presence of two such sites immediately upstream of the 5' hypersensitive sites and downstream of the 3' hypersensitive site. However, they found a number of other sites in the β-globin domain and it is presently not clear which of these are functionally important. The DCR might also contain one or several enhancer-like sequences, which can exert their effect over very large distances (>50 kb), without stimulating the most proximal promoters in the foetal and adult stages. Interestingly, a similar control region has recently been identified for a T-cell specific gene CD2, albeit at the 3' side of that gene.[19]

At least four different control regions in the β-globin gene have been identified: the DCR, the promoter and two enhancer sequences and at least two in the γ-globin gene, the DCR and the promoter. On the basis of this and the in-vivo mutations, we propose the following model for the control of globin gene expression during development. The dominant control region determines the activity of the locus. In inactive non-erythroid tissues there are no hypersensitive sites and the genes are not accessible to *trans*-acting factors. In erythroid cells the DCR becomes hypersensitive (possibly not requiring replication[20]) and renders the chromatin accessible to factors. The region made accessible may be delineated by nuclear matrix binding sites. Action of the dominant control region then mediates the binding of *trans*-acting factors to regulatory promoter and enhancer sequences immediately surrounding the globin genes. The latter process largely determines the stage-specific expression and in the case of the β-globin gene, this process would involve possible negative regulatory factors and sequences to suppress the gene at early stages and positive factors acting on the enhancers and promoter to set up an extremely active transcription complex in adult stages.

Obviously, the discovery of these sequences has a number of practical consequences. Firstly, the ability to express foreign genes under the control of the human β-globin DCR, will allow a high level of expression of foreign genes and their products in erythroid cell lineages. The data obtained with the α-globin, Thy-1 and tk-neo genes in such experiments indicate that the rate limiting step at the transcriptional level will be determined by the efficiency of the promoter, i.e. expression of the Thy-1 gene is much lower than that of the α-globin and the tk-neo gene.[12] Potential low expression problems could be overcome by using the β-globin promoter in hybrid gene constructs, after which the downstream processing and RNA stability would become the limiting factors. The second interesting application of these sequences would be their use in gene therapy. Inclusion of these sequences in retroviral vectors would solve the low expression problems observed in bone marrow stem cell transformation and transplantation experiments.[21] At present the authors are testing this possibility by determining the smallest possible, fully functional, β-globin gene and DCR for inclusion in retroviral vectors.

ACKNOWLEDGEMENT

The authors are grateful to Cora O'Carroll for typing this manuscript.

REFERENCES

1. van der Ploeg, L. H. T., Konings, M., Oort, D., Roos, L., Bernirti, L. & Flavell, R. A., Gamma-β-thalassaemia studies showing that deletion of the gamma- and delta-genes influences β-globin gene expression in man. *Nature,* **283** (1980) 637–42.
2. Vanin, E. F., Henthorn, P. S., Kioussis, D., Grosveld, F. & Smithies, O. Unexpected relationships between four large deletions in the human β-globin gene cluster. *Cell,* **35** (1983) 701–9.

3. Kioussis, D., Vanin, E., deLange, T., Flavell, R. A. & Grosveld, F., β-Globin gene inactivation by DNA translocation in gamma β-thalassaemia. *Nature*, **306** (1983) 662–6.

4. Wright, S., Rosenthal, A., Flavell, R. A. & Grosveld, F. G., DNA sequences required for regulated expression of β-globin genes in murine erythroleukemia cells. *Cell*, **38** (1984) 265–73.

5. Taramelli, R., Kioussis, D., Vanin, E., Bartram, K., Groffen, J., Hurst, J. & Grosveld, F., *Nucl. Acids Res.*, **137** (1986) 2088–92.

6. Tuan, P., Solomon, W., Qiliang, L. & Irving, M., The 'β-like-globin' gene domain in human erythroid cells. *Proc. Nat. Acad. Sci., USA*, **8** (1988) 6384–8.

7. Forrester, W., Takegawa, S., Papayannopoulou, T., Stamatoyannopoulos, G. & Groudine, M., Evidence for a locus activation region: The formation of developmentally stable hypersensitive sites in globin-expressing hybrids. *Nucl. Acids Res.*, **15** (1987) 10159–77.

8. Grosveld, F., Blom van Assendelft, G., Greaves, D. & Kollias, G., Ectopic expression of Thy-1 in the kidneys of transgenic mice induces functional and proliferative abnormalities. *Cell*, **51** (1987) 21–31.

9. Magram, J., Chada, K. & Costantini, F., Developmental regulation of a cloned adult β-globin gene in transgenic mice. *Nature*, **315** (1985) 338–40.

10. Townes, T., Lingrel, J., Chen, H., Brinster, R. & Palmiter, R., Erythroid-specific expression of human β-globin genes in transgenic mice. *EMBO J.*, **4** (1985) 1715–23.

11. Kollias, G., Wrighton, N., Hurst, J. & Grosveld, F., Regulated expression of human gamma-, β-, and hybrid gamma β-globin genes in transgenic mice: manipulation of the developmental expression patterns. *Cell*, **46** (1986) 89–94.

12. Blom van Assendelft, M., Hanscombe, O., Grosveld, F. & Greaves, D. R., The beta globin dominant control region activates homologous and heterologous promoters in a tissue-specific manner. *Cell*, **56** (1989) 969–77.

13. Nandi, A., Roginski, R., Gregg, R., Smithies, O. & Shoultchi, A., Regulated expression of genes inserted at the human chromosomal β-globin locus by homologous recombination. *Proc. Nat. Acad. Sci., USA*, **85** (1988) 3845–9.

14. Talbot, D., Collis, P., Antoniou, M., Vidal, M., Grosveld, F. & Greaves, D. R., A dominant control region from the human beta globin focus conferring integration site independent gene expression. *Nature*, **338** (1989) 352–5.

15. Lawson, G., Knoll, B., March, C., Woo, S., Tsai, M. & O'Malley, B., Definition of 5′ and 3′ structural boundaries of the chromatin domain containing the ovalbumin multigene family. *J. Biol. Chem.*, **257** (1982) 1501–7.

16. Stadler, J., Larsen, A., Engel, J., Dolan, M., Groudine, M. & Weintraub, H., Tissue specific DNA cleavages in the globin chromatin domain introduced by DNA ase1. *Cell*, **20** (1980) 451–60.

17. Gasser, S. & Laemmli, U., Cohabitation of scaffold binding regions with upstream/enhancer elements of three developmentally regulated genes. *Cell*, **46** (1986) 521–30.

18. Harman, A. & Higgs, D., Nuclear scaffold attachment sites in the human globin gene complexes. *EMBO J.*, **7** (1988) 3337–44.

19. Greaves, D. R., Wilson, F. D., Lang, G. & Kioussis, D., Human CD2 3′-flanking sequences confer high-level T cell-specific position independent gene expression in transgenic mice. *Cell*, **56** (1989) 979–86.

20. Baron, M. & Maniatis, T., Rapid reprogramming of globin gene expression in transient heterokaryons. *Cell,* **46** (1986) 591–602.
21. Dzierzak, E., Papayannopoulos, T. & Mulligan, R. Lineage-specific expression of a human beta-globin gene in murine bone marrow transplant recipients reconstituted with retrovirus transduced stem cell. *Nature,* **331** (1988) 35.

Chapter 11

THE PRODUCTION OF PHARMACEUTICAL PROTEINS
IN THE MILK OF TRANSGENIC ANIMALS

ALAN E. SMITH

Integrated Genetics, Inc., Framingham, Massachusetts, USA

INTRODUCTION

Proteins can be made by a variety of routes for applications ranging from industrial enzymes to human therapeutics. Recently interest in this subject has increased greatly since the ability to manipulate DNA and to transfer it into different host cells has meant that virtually unlimited amounts of individual proteins can now be made.

Two central issues in the production of proteins are quality and cost. Because proteins are complex molecules with molecular weights in the tens of thousands, they are readily subject to denaturation and inactivation. Equally important, the amino acid backbone of many proteins is chemically modified. The best known of such modifications are addition(s) of carbohydrate moieties, but many others such as acyl, acetyl, phosphate, sulphate and carboxyl groups are known to occur naturally on different proteins in different tissues. These modifications are important since in many cases they influence the biological activity of the protein. Because such modifications do not occur universally, particular attention has to be paid to the ability of any chosen host cell to bring about the modifications required for full activity of a given protein of interest.

In general, bacterial, yeast and insect cells do not modify proteins in exactly the same way as mammalian cells. This is best documented in the case of carbohydrate addition where such modification is either largely absent (bacteria) or grossly different (yeast and insect cells). Consequently, considerable effort has gone into development of mammalian cell systems that can express introduced genes so that proteins for human use can be made that mimic exactly their natural counterparts. The two best known mammalian systems are the Chinese hamster ovary (CHO) cell containing integrated, amplifiable vectors and the mouse C127 cell system containing episomal or integrated bovine papilloma virus (BPV) DNA based vectors. These are described elsewhere in this volume. There are a number of differences between these two systems but in general both are able to produce proteins in the 1–20 mg/litre/day range, both are scalable to thousands of litres and both produce accurately modified proteins which in some cases have been shown to be active in human clinical trials. These systems clearly establish the feasibility of producing proteins in mammalian cell culture and although CHO and C127 cells are not capable of all modifications, it is likely that as special requirements arise for particular proteins appropriate new host cells will be developed.

One problem remains; the culturing of mammalian cells is expensive. For high value human pharmaceuticals where only a small dose (tens of micrograms) is required this will not present a problem. However, for many pharmaceutical proteins, and for proteins for industrial use, bacterial fermentation will produce inactive protein and mammalian cells will be prohibitively expensive.

TRANSGENIC ANIMALS FOR PROTEIN PRODUCTION

The production of proteins in the bodily fluids of transgenic animals presents another alternative for the production of modified proteins. To be competitive, this technology would need to modify proteins accurately, to be inexpensive, and to result in a high yield of biologically active product. That these criteria are likely to be met has been established using transgenic mice as a model.

Microinjection

Transgenic animals are produced by introduction of foreign DNA into the germ line of the recipient animal. Following injection into the fertilized egg and reimplantation into foster mothers, the resulting offspring are tested for the presence of the introduced DNA. If present, the transgenic animal is tested for the appropriate phenotype. In this way, many transgenic lines have been established.[1] Key elements in the DNA introduced into a transgenic animal are the sequences coding for the protein of interest, and adjacent segments needed to control the time, place and extent of expression of the introduced DNA (Fig. 1).

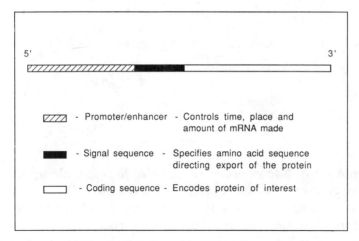

Fig. 1. Elements of introduced gene.

Vector Design

Much research in transgenic animals attempts to establish the best configuration of DNA to achieve the desired phenotype in the transgenic animal. This subject is discussed in detail elsewhere in this volume (Chapter 10). Although much early work utilized cDNAs and produced appropriate transgenic animals, it seems likely (though is not yet established) that genomic DNA including the naturally occurring introns gives higher levels of transcription. It is also not yet established which promotors/enhancers work best. So far, it has been possible to get correctly regulated expression in at least some transgenic animals using surprisingly small segments of upstream control DNA, including in some cases DNA from different species. However, it is possible that very large segments of DNA from regions adjacent to the immediate control elements, or the newly described 'dominant control region' (DCR) will give greater numbers of useful transgenics and give better expression levels (see Chapter 10).

Selected Tissue

For the commercial production of proteins, the ideal tissue for expression would be capable of making large amounts of modified protein in some readily accessible fluid devoid of other proteins and physically sequestered from the host animal. Although no such tissue exists, the mammary gland comes close. Milk is extremely abundant in many domesticated species. It is very inexpensive to produce and it contains

Table 1
Protein Yield in Milk

Composition of milk	
Total protein	29 g/litre
Total caesin	25 g/litre
Amount of milk	
Cow	30 litres/day
Goat	3 litres/day

large amounts of a few species of protein; for example up to 30 g/litre mostly as caesins in cows, sheep and goats (Table 1).

Promoters and enhancers for mammary gland specific expression have been identified, so that the major issues in establishing the feasibility of this route of protein production became (a) what levels of expression could be achieved, (b) would the promoter be sufficiently tissue specific to cause no health problems to the host animal, (c) would the protein be excreted into the milk, (d) would the protein be stable in the milk, and (e) would the protein be biologically active?

t-PA EXPRESSION IN TRANSGENIC MICE

The author's work on the feasibility of protein production in transgenic milk used human tissue plasminogen activator (t-PA) as the model protein.[2] It is an approved drug for human use, it is expensive to produce, its price is subject to competitive pressures, it is modified by both proteolytic cleavage and by glycosylation and, importantly, there are simple and sensitive assays for its presence.

A human cDNA for t-PA was attached to the 5′ upstream region from the mouse whey acidic protein gene. This gene encodes an abundant protein in mouse milk and previously had been partially characterized and its expression shown to be tissue specific.[3] The signal sequence of human t-PA was used to specify export from the cell, making the assumption that transfer of a nascent protein into the lumen of endoplasmic reticulum is universal, requiring only one of many such alternative hydrophobic sequences.

Seven transgenic animals were produced. These contained widely varying numbers of integrated copies of the human t-PA cDNA. Upon appropriate breeding from the founder mice, protein levels in the milk of lactating transgenic mice was found to vary widely and to bear no relation to the cDNA copy number. Indeed, the highest level of expression was found in an animal likely containing only a single copy of human t-PA DNA. This line has been bred successfully over many generations. The animals live an apparently normal life span, are able to support young in spite of the human protease in the milk, and suffer no detectable health problems. It is assumed that the high expression level in this particular mouse line reflects the integration of the t-PA cDNA into a position in the chromosome favourable to transcription of (mammary gland) genes.

The t-PA in the milk has been shown to be enzymatically active using a semi-quantitative fibrin clot lysis assay. The levels of protein were measured by ELISA to be about 50–100 mg/litre. Polyarylamide gel electrophoresis showed the protein to be predominantly two chain t-PA and to migrate with approximately the same mobility as human t-PA derived from a tumour cell line.[4] The high level of protein production is extremely encouraging and exceeds by as much as tenfold the amount of protein made in a large-scale bioreactor. Since only seven animals were tested, the expectation is that by examination of larger numbers this level could be exceeded.

Two major avenues of development now require attention. Can this result be translated to a higher-milk-producing species such as the cow, and can expression levels be increased by improved design of vectors?

HIGHER ANIMALS

There are a large number of factors to be considered in deciding which animal species to examine for the feasibility of large scale protein production in transgenic milk (see Table 2).

Cows are an obvious choice but for initial work they suffer the great disadvantage, in comparison with smaller animals, of a very long gestation period and time to sexual maturity. Other factors such as milk yield, protein content of milk, knowledge of the appropriate molecular biology of mammary gland gene expression, biochemistry of milk and knowledge of and ease of embryological manipulation all need consideration. Upon such analysis, two species emerge as attractive candidates — sheep and goats.

Table 2
Factors to be Considered for Transgenic Milk Production

Easily handled animal	Rapid gestation and maturity
Good milk yield	Molecular biology
Embryology available	Milk biochemistry

Integrated Genetics and Tufts University School of Medicine Veterinary Department have established a collaboration to develop a herd of transgenic goats, in the first instance producing a second generation t-PA molecule called longer acting t-PA.[5] The gestation period of goats is 5 months and the time to sexual maturity about the same. It takes some time therefore to obtain data with which to estimate the likely expression levels in goats. Another group has produced transgenic sheep, but so far production levels of protein in their milk are low.[6]

Vectors for Larger Animals

Major questions concerning the choice of vector include which animal species of DNA and which promoter/enhancer combination should be used, and should small cDNA containing vectors or large macro vectors including genomic coding sequences and surrounding DNA be used? Experiments addressing these issues are underway in a number of laboratories including our own. This work is limited by the time and expense of breeding larger animals and by the uncertain validity of extrapolating results obtained in mice to other species. Examination of very large numbers of transgenic animals for the few with very high expression levels based fortuitously on a favourable site of integration is a possible alternative to extrapolating from mouse models but is very expensive when working with larger animals.

Remaining Scientific Issues

The studies outlined above are likely to produce transgenic animals capable of yielding large quantities of milk containing the protein(s) of interest. Important issues remain; will the proteins be of good quality, and will they be extractable from milk in good condition and reasonable yield?

It is well established that the mammary gland is capable of a wide range of protein modifications. Milk proteins are extensively phosphorylated for example, and our feasibility studies in transgenic mouse milk show that the t-PA is glycosylated. It is likely however, that some specialized modifications will not take place. In this case, and if the protein is otherwise of good quality, an enzymatic means to add a final modification might prove possible. It would not be surprising if carbohydrate addition in transgenic mouse mammary gland differ(s) slightly from that in engineered mouse C127 cells. Indeed there is good evidence for differences in carbohydrate even between different mammalian cells, between different growth conditions and between different modes of production. These differences, while of concern need not cause undue problems if they can be controlled (that is if the protein product can be made reproducibly) and if it can be shown that the protein is biologically active. The likelihood of proteins made by any recombinant route being absolutely identical to their naturally occurring human counterparts is low. For this reason, it is likely that for pharmaceutical use reproducibility, efficacy and safety will become the paramount criteria, not necessarily absolute identity.

Whether a given protein will be stable in milk is impossible to answer theoretically but it is simple to test by mixing experiments. In the case of tPA, the protein produced in milk is two chain rather than single chain. The protease in milk responsible for the cleavage is unknown but since two-chain material is biologically active, this form of the molecule should be satisfactory.

Purification of proteins from milk is not expected to pose major technical hurdles, partly because the starting material will be abundant, the pharmaceutical protein itself will be present in relatively large amounts and the composition of milk is known to be relatively simple. Milk fractionation schemes already exist and it should be straightforward to modify such procedures to purify desired proteins. As with the production of any human pharmaceutical, the conditions for purification will need to comply with good manufacturing practices and special attention will have to be paid to eliminating trace contaminants of milk proteins and DNA and to inactivating any biological agents that could contaminate the product. In this regard, however, purification from transgenic milk differs little from purification from conditioned medium from engineered mammalian cells where these same issues arise.

While these issues are not trivial and require time and experience to address with confidence, none seem to present more than the usual developmental problems associated with introducing a modified production technology.

REGULATORY ISSUES

Several regulatory issues relating to the production of human pharmaceuticals have already been mentioned in relation to purification and specifications. Maintaining the transgenic animal herd in a clean environment will be essential, but in the light of the potential value of the animals, close attention to their well-being and health would seem appropriate even in the absence of regulatory requirements.

Another factor is that animal-derived protein products have been used successfully and safely in the past. Insulin is an example of a protein which has been purified from animal tissue for decades, and several vaccines have been produced in eggs. In the regulatory sense then, protein products from animals is not a totally new area.

The author anticipates that a number of items will require debate before a full regulatory policy on transgenic animals for pharmaceutical protein production emerges. For example, would production need to be from a single transgenic line or could the protein from several lines be mixed? Again, however, such issues do not appear to be hurdles likely to prevent commercialization of this technology.

PATENTING

Perhaps surprisingly one of the most contentious issues arising from the development of transgenic animals is the question of whether such animals are patentable. US law on this subject states clearly that 'anything under the sun that is man-made is patentable'. That law, however, has been challenged in relation to living organisms mostly on the grounds that the Supreme Court pivotal yet controversial Chakrabarty decision was rendered in favour only by a close, split decision. Strong pressure groups have proposed that patenting of transgenic animals should be subject to a moratorium while a change in the law is considered. One major objection to patenting animals comes from farmers, who are concerned at the prospect of having to pay royalties and/or licence fees for animals bred from the original transgenic animal, designed, for example, for improved meat production or disease resistance. Other objectors focus on religious, moral and ethical issues or on the environmental impact of transgenic animals.

The moral/ethical/religious arguments revolve around such matters as sanctity of life, interference with the breeding of animals, and interference with the stewardship of the environment given to humankind by the Creator. These of course are issues on which strong and genuine views are held, but it is difficult to see why they apply specifically to transgenic animals rather than farming, animal domestication and breeding in general.

The environmental issue, at least as applied to animals for protein production, is largely a matter of misunderstanding in that if successful very small numbers of transgenic animals will be all that is required to provide the world supply of many pharmaceuticals. The likelihood of those animals escaping and having an environmental impact is negligible. If large numbers of animals are required, production would not be economical, and the more widespread potential problem would not arise.

A US government subcommittee, under Congressman Kastenmeier, has reported on the subject of patenting animals.[7] The report concluded that there is a special case to be made for farmers and the patenting of farm animals but that patenting of transgenic animals for models of human diseases and for protein production should go ahead as presently allowed by law. Indeed in the interim, a patent was awarded to Harvard University for transgenic mice containing human oncogenes.

The subject of patenting is of some relevance to the science of transgenic animals in that in the absence of patents, the likelihood of investment to develop the technology is reduced. Although for the moment the question of patents appears to have been resolved, it is of some concern that the coalition against patenting has pledged continued opposition in the next Congress.

ECONOMICS

The reason for developing transgenic animals for protein production in milk was the belief that this may be the best way to produce some proteins most economically. The potential for cost reduction is enormous. The protein concentration in milk is tens of grams per litre and of protein in conditioned media from bioreactors is at best tens of milligrams per litre per day. The early feasibility studies in mice show that production of 100 mg/litre of a foreign protein in milk is achievable relatively easily. It is therefore not unrealistic to think that an increase in protein concentration of one hundred fold could be achieved. In addition, the cost of conditioned media coming from a plant producing protein from engineered mammalian cells can be as high as a hundred dollars/litre whereas the cost of feeding animals results in milk output at less than one dollar/litre. Even allowing for the cost of producing the transgenic animals (which incidentally compares favourably with building a dedicated bioreactor plant), and for the likely lower yields resulting from purification from milk, very substantial savings should be possible by utilizing the transgenic route.

Not all proteins however, are obvious candidates for production from transgenic animals. Highly potent molecules requiring only very small doses are economical when produced by more traditional routes. Examples include growth factors such as erythropoeitin. In other cases when second and third generation molecules are anticipated, the time lag from improved molecule to large scale transgenic production might be too long. Nevertheless, for pharmaceuticals of high or intermediate value but which require a high dose, or for lower value proteins, transgenic animals may prove the production means of choice.

ACKNOWLEDGEMENTS

The author thanks his many colleagues at Integrated Genetics and the National Institutes of Health, particularly Katie Gordon, Heiner Westphal, Lothar Hennighausen, Gordon Moore and Suzanne Groet for many fruitful discussions on this subject and Ms Mary Lewis for help with the manuscript.

REFERENCES

1. Palmiter, R. D. & Brinster, R. L., Germ-line transformation of mice. *Ann. Rev. Genet.*, **20** (1986) 465–99.
2. Gordon, K., Lee, E., Vitale, J. A., Smith, A. E., Westphal, H. & Hennighausen, L., Production of human tissue plasminogen activator in transgenic mouse milk. *Biotechnology,* **5** (1987) 1183–7.
3. Andres, A. C., Schonenberger, C. A., Groner, B., Hennighausen, L., LeMeur, M. & Gerlinger, P., Ha-ras oncogene expression directed by a milk protein gene promoter: tissue specificity, hormonal regulation, and tumor induction in transgenic mice. *Proc. Nat. Acad. Sci., USA,* **84** (1987) 1299–303.
4. Pittius, C. W., Hennighausen, L., Lee, E., Westphal, H., Nicols, E., Vitale, J. & Gordon, K., A milk protein gene promoter directs the expression of human tissue plasminogen activator cDNA to the mammary gland in transgenic mice. *Proc. Nat. Acad. Sci., USA,* **85** (1988) 5874–8.
5. Lau, D., Kuzma, G., Wei, C. M., Livingston, D. & Hsuing, N., A modified human tissue plasminogen activator with extended half-life *in vivo. Biotechnology,* **5** (1987) 953–8.
6. Clark, A. J., Simons, J. P., Wilmut, I. & Lathe, R., Pharmaceuticals from transgenic livestock. *Trends Biotechnol.,* **5** (1987) 20–4.
7. Patents and the Constitution: Transgenic Animals 1987, Hearings before the Subcommittee on Courts, Civil Liberties and the administration of justice of the Committee on the Judiciary, House of Representatives, One Hundredth Congress, Serial No. 23, US Government Printing Office, Washington, D.C.

Chapter 12

THE PRODUCTION OF USEFUL PROTEINS FROM TRANSGENIC PLANTS

Michael Bevan & Gabriel Iturriaga

IPSR Cambridge Laboratory, Trumpington, Cambridge, UK

INTRODUCTION

In many respects this chapter can be seen as a statement of the obvious — mankind has been obtaining tons of important protein products from plants for industrial uses for generations. What genetic engineering can do is to add a new source of variation to the natural variation that is already exploited in agriculture. In this review we will outline the recent progress made in the understanding of gene expression, gene product targetting, gene isolation and tissue culture that will be important for the development of plants and plant tissue culture as sources of novel proteins.

IMPORTANT METHODS

Gene Transfer to Higher Plants

Several methods are currently used for introducing DNA into higher plants. Many Solanaceae and Brassiceae have sufficiently well described tissue culture so that direct DNA uptake into protoplasts, followed by protoplast regeneration into fertile plants, can be used. Also, *Agrobacterium tumefaciens* and *Agrobacterium rhizogenes* vectors are widely used, coupled to methods that allow for the rapid regeneration of fertile plants from transformed cells. Important crop plants in this group include potato, tomato, rape, soybean, cucumber, carrot, alfalfa, lettuce, poplar, sugar beet and cotton, although the methods are not routine for many of these species. Plants that have less well understood tissue culture systems, such as many of the major grain crops (wheat, rice, maize) are the subject of intensive research aimed at transformation using different methods. These methods include introducing DNA into protoplasts and regenerating plants from these transformed cells, microinjection of various plant organs such as the egg or pollen, or ballistic delivery of DNA[1] to meristematic zones or pollen. Currently there is much excitement surrounding the latter technique, as it promises to circumvent laborious tissue culture steps. It is thought that methods will be found for obtaining stable, heritable DNA introduction and gene expression, in the few remaining recalcitrant species within the next 3 years.

Gene Expression in Transformed Plants

In this section the factors that contribute to maximal and tissue specific gene expression in transgenic plants are outlined. For many of the traits currently being engineered into plants, low levels of expression limit their application, hence current approaches to this problem have been emphasised.

(a) Transcriptional Control

Initial experiments in gene expression in transformed plants used promoters of non-plant origin, for example, those from *Agrobacterium* Ti plasmid genes such as nopaline synthase and octopine synthase, and from DNA viruses such as Cauliflower Mosaic Virus (CaMV).[2] These were found to be expressed in most tissue types and were called constitutive, and they and other related promoters still form the 'workhorse' promoters for expressing resistance markers. Having used these promoters to construct vectors, it was then possible to study putative transcriptional regulatory sequences from plant genes. Among the first studied were those from nuclear genes encoding components of the photosynthetic apparatus such as the small subunit of Rubisco (ribulose biphosphate carboxylase) and chlorophyll a/b binding protein.[3] These genes encode abundant protein products, but they are members of gene families with many members in diploid species so the expectation that a single promoter could direct the expression of an extremely abundant RNA was not fulfilled. However, in the case of the small subunit gene family in *Petunia*, one member encodes approximately half of the total mRNA. Surprisingly the expression of nearly all genes encoding proteins of agricultural importance have used the CaMV 35S promoter, which encodes an abundant RNA. This promoter has recently been shown to exhibit preferential activity in the cells of the phloem,[4] that is, it is expressed most in highly vascularised organs such as stems and roots. Nevertheless it has reasonable expression in most other cell types. Multimerisation of the upstream activation sequences increases the amount of steady state transcript in transient assay systems substantially,[5] but this effect is not dramatic in stable transformants. Table 1 summarises the cell specificity and inducer (where known) of some of the plant promoters isolated to date.

(b) mRNA Stability and Translatability

The inclusion of an intron from a maize ADH1 gene in either the coding or non-coding region of a gene significantly increased protein production in transient assays.[6] This stimulation was specific for monocot cell lines, and it was thought that this could either reflect a monocot- and dicot-specific splicing mechanism (which has been observed), a monocot specific enhancer in the intron, or a combination of both of these factors. Clearly in order to maximise protein expression the inclusion of an intron is of potential importance, although the mechanisms of action are not yet understood and may not be general.

The wealth of data generated from studies of plant RNA viruses has allowed for a thorough analysis of the factors that influence translation in plants. This has led to the incorporation of 5′ leader sequences from two viruses, tobacco mosaic virus (TMV), and alfalfa mosaic virus (AlMV) into chimaeric gene constructs, as they are translated very efficiently *in planta* and *in vitro*. Stimulation of protein expression by 5–10

Table 1

Promoter/gene	Cell specificity	Inducer
Rubisco small subunit	Photosynthetic	Light
Chlorophyll a/b binding protein	Photosynthetic	Light
Chalcone synthase	Not tested	Light (UV b)
Alcohol dehydrogenase	Not tested	Anaerobiosis
Heat shock (p18 and p70)	Not tested	Heat shock
Proteinase inhibitor	Not tested	Wounding, pathogens
Win2, chitinases	Most cell types	Wounding, pathogens
Phenylalanine ammonia-lyase	Xylem, epidermis	Wounding, pathogens
Chalcone isomerase	Anthers	Not known
CaMV 35S	Predominantly phloem	Not known
Soybean conglycinin	Embryo	Not known
Rape napin	Embryo	Abscisic acid
Wheat glutenin	Endosperm	Not known
Maize zein	Endosperm	Not known
Bean phaseolin	Embryo	Not known
Potato patatin	Most cell types	Sucrose

fold is observed in transient assay of genes containing these 'translational enhancers',[7] but there are currently no data on expression in stably transformed plants. However, there is every likelihood that this effect will also be seen in that situation. It is thought that these sequences act by decreasing the secondary structure at the 5' end of the mRNA, thus allowing avid ribosome binding. This information has also been useful in dictating the structure at the 5' end of constructed genes — remnants of restriction sites, especially those that are GC rich, form hairpin loops that are thought to either interfere with ribosome binding or impede ribosome movement on the mRNA.

Potential differences in codon usage between plants and other organisms have not yet been investigated in attempts to maximise protein expression. It is possible that there may be codons recognising tRNAs that are rate limiting in plants, and therefore the identification of these and their mutagenesis to recognition of abundant plant isoacceptors might be worthwhile.

(c) Targetting to the Endomembrane System

Among the most abundant proteins in plants are the storage proteins of somatic and embryonic tissues. These are synthesised from abundant transcripts and accumulate in specialised vacuoles, where they form stable complexes, often composed of different proteins.[8] It is thought that factors in addition to abundant and stable mRNAs contribute to this large accumulation of protein. Principal among these is the sequestration of proteins in membrane bound organelles. This may decrease the degradation of proteins that normally occurs in the cytoplasm and additionally provide an environment for assembly into stable complexes, which may provide further protection from degradation. Finally, modifications such as N-linked glycosylation, which occurs during targetting to the endomembrane system, may also contribute to increased stability. It may also be feasible, utilising the appropriate expression system, to fuse the coding region of unstable proteins to more stable storage proteins in order to enhance stability.

Little is known about targetting of non-plant proteins to the endomembrane system in higher plants. Recently the putative amino-terminal transit sequence from a potato tuber storage protein called patatin has been used to target a cytosolic bacterial enzyme (β glucuronidase, GUS) to the endomembrane system of transgenic tobacco.[9] The GUS protein was glycosylated at a cryptic site(s), and after tunicamycin treatment active GUS was able to be recovered from the cells. This indicated that enzymatically active tetramers were able to assemble in the endomembrane system. In addition, it was thought that

GUS targetted to the endomembrane system was more stable than cytoplasmic GUS. If substantiated, this observation has clear implications for engineering high levels of protein in transgenic plants. A consideration related to targetting to the endomembrane system is that of post-translational modifications to proteins. Plants elaborate complex carbohydrates on proteins much as animal and yeast cells do, and this ability could be exploited. Finally, it is important to consider the possible secretion of engineered proteins from plant cells. One could envisage immobilised plant cells in a reactor secreting protein into the medium, thereby lowering the amount of downstream processing required for purification. Two factors may make this scenario less attractive however. First, the cell wall may impede or bind certain secreted proteins, as it is negatively charged. Secondly, plant cells in culture secrete massive amounts of polysaccharide that would be a problem during downstream processing. Nevertheless, secretion may be desirable for particular combinations of protein and plant cell.

(d) Organ Specificity of Expression

If foreign protein production in plants is to exploit fully the high levels of efficiency obtained by modern agricultural practises in generating biomass from sunlight and soil, then it would be reasonable that the production and harvesting of foreign proteins be integrated into normal agricultural production. The way that genetic engineering can address these issues is to obtain foreign protein expression in organs of particular plants that can be harvested using available techniques. For example, expression in grains of wheat or barley, or the seeds of rape, would allow rapid and efficient initial 'purification' of the expressed protein using combine harvesters. Also, expression in roots of sugar beet, or potato tubers, would integrate harvesting with production. In the case of sugar beet, the availability of factories for extraction from sugar beet on a massive scale may be economically advantageous, particularly as these plants require little modification for cold extraction of proteins, and they lie idle for part of the year. Similarly, the extraction of starch from potato tubers occurs by a cold extraction method that would allow for the recovery of native protein on a huge scale. It can be seen that although only a few percent of total plant protein can currently be synthesised from a transgene, with the appropriate tissue and organ specificity, coupled to efficient purification methods, plant production could theoretically yield enormous amounts of foreign protein from quite small fields. Table 1 contains information on organ and tissue specific promoters that may be used to direct expression in organs customarily harvested. As discussed previously, storage organs generally contain classes of abundant proteins; many of the genes encoding these proteins have been characterised and could be used in the near future.

APPLICATIONS OF GENETIC ENGINEERING TO CROP IMPROVEMENT

Much of the impetus provided for the rapid development seen in plant molecular biology in the last 5 years has come from the realisation by many companies involved in agriculture that gene transfer techniques could widen the application of current technologies, e.g. herbicide production, and also open up entirely new avenues for exploitation. The progress in this field by companies committed to agricultural bio-technology has been dramatic, and we will review the progress made in this field in this section.

(a) Herbicide Resistance

Several different approaches have been taken to introduce genes that potentially confer increased resistance to herbicides. Resistance to the herbicide glyphosate has been engineered using two different strategies. First, an endogenous plant gene encoding EPSP (enoyl-pyruvyl-shikimate phosphate) synthase, the target enzyme inhibited by glyphosate, was overproduced to confer resistance.[10] EPSP synthase, like ALS (acetolactate synthase) which will be discussed later, are involved in branched chain amino acid biosynthesis, which occurs by a pathway unique to higher plants, hence the very low toxicity of these compounds to other organisms. Subsequent work has used mutants of the plant gene, selected in bacteria for enhanced resistance to glyphosate, to obtain resistance to field doses. One of the first field trials of a genetically engineered plant was to test the efficacy of glyphosate resistance in transgenic tomatoes. A second approach used a gene encoding EPSP synthase from *Salmonella*, which was insensitive to glyphosate, in transgenics.[11] This again provided useful levels of resistance to the herbicide. Current work in expressing EPSP synthase is aimed at getting even higher levels of expression in particular regions of the plant known to accumulate glyphosate. Sulphonylurea herbicides inhibit ALS, and genes encoding mutant versions of this enzyme that conferred resistance in tobacco were isolated, along with the wild-type gene. Mutations were made in the wild-type gene that were the same as some of those found in the mutant genes, and when the new mutants were transferred to plants they conferred very high levels of resistance.[12] Similarly, mutant ALS genes from *Arabidopsis* selected for resistance to sulphonylureas were also able to make transgenic plants resistant to field doses of herbicide.[13] A different strategy has been taken with engineering resistance to two other herbicides. Genes from bacteria encoding enzymes that detoxify herbicides were introduced into plants under the control of plant promoters, and these have been very successful. Nitrilase from the soil bacterium *Klebsiella ozaenae* destroys bromoxynil, a photosystem 2 inhibitor. When expressed in transgenic tobacco, it made plants resistant to large doses of bromoxynil.[14] A gene encoding phosphinothricin acetyl transferase from *Streptomyces*, when expressed in plants, confers resistance to high levels of the herbicide Basta, or phosphinothricin.[15] This chemical inhibits glutamine synthetase, and the plants die due to ammonia accumulation. Finally, a 2,4-D monoxygenase from a bacteria has been expressed in plants and this confers resistance to 2,4-D (Willmitzer, L., personal communication). Potential regulatory problems may limit the general application of detoxification strategies to agriculture, as the detoxified residues must also be examined for environmental impact, at considerable cost.

(b) Insect Resistance

The most successful approach has utilised the *Bacillus thuringiensis* crystal toxin (Bt toxin), which specifically paralyses the gut of insects. There are a variety of strains that exhibit markedly different toxicities towards Lepidopteran, Dipteran and Coleopteran pests. The protein is extremely specific and toxic, and only very low levels of expression were needed to inhibit feeding on expressing plants.[16] In fact the mRNA for Bt toxin is very unstable in plants, and is barely detectable even when expressed from highly active promoters such as CaMV 35S. Using this strategy it is hoped to engineer resistance to a wide variety of insect pests that currently do much damage to crops such as maize, cotton and potatoes. There are two significant advantages to the use of Bt insecticidal toxins. First, they have been in use for many years and no significant resistance has developed to them. Secondly, the widespread use of Bt toxins as topical applications means that much of the necessary knowledge on the effect of the protein on the ecology of different areas is already known. A second approach to insect resistance uses endogenous plant genes known to inhibit insect digestion. A trypsin inhibitor from cowpea had been shown to inhibit the feeding of brucid beetles, and when expressed in leaves of tobacco plants (it is normally found in large amounts in seeds) at levels of 2% of total protein using a CaMV 35S promoter, a significant reduction in feeding and larval death was observed.[17] Similar experiments have also been started using a proteinase inhibitor gene from potato.

(c) Virus Resistance

Two different approaches to engineering resistance to RNA viruses in crop plants have been undertaken. They both exploit the principle of cross protection. The prior expression of a virus in a plant can often confer resistance to superinfection by a related virus, and this phenomenon has been called cross protection. Cross protection has been engineered by expressing viral components as nuclear genes. For example, the coat protein of several important plant pests such as tobacco mosaic virus (TMV) protects against TMV infections to agriculturally significant levels in tomato.[18] Similarly, protection against potato viruses X and Y has been achieved in potato.[19] The mechanism of coat protein mediated cross protection is still obscure, but a possible mechanism may involve the inhibition of unpackaging of the virus. For success, high levels of coat protein (up to 1% of total protein) have to be expressed from the nuclear gene. The CaMV 35S promoter has been most useful for obtaining these levels in the appropriate cell types. Plant RNA viral infections are often accompanied by characteristic satellite RNAs that can dramatically alter symptom production. The expression of a satellite sequence as a nuclear transcript attenuates the replication of cucumber mosaic virus in tobacco, possibly by competing for limiting replication components.[20]

Antisense RNA has been rather disappointing in attentuating virus growth in plants, possibly because of the very high levels of viral RNA that accumulate in infected cells. It is also possible that antisense RNA exerts its effect by inhibiting mRNA maturation in the nucleus, and has no role to play in the cytoplasm, such as inhibiting translation. There are important classes of plant DNA viruses that are particularly destructive pests, such as the Geminiviruses, which may be more suitable targets for antisense inhibition.

(d) Alterations in Metabolism

Gene transfer can be used to alter the activity of endogenous genes in predictable ways, for example, by replacing an endogenous gene with a mutated version, or by antisense inhibition. As gene replacement is not a possibility in higher plants at the moment, the latter method has been used to down regulate endogenous genes. The expression of poly-galacturonase during fruit ripening was able to be inhibited by greater than 90% in transgenic tomatoes using antisense polygalacturonase.[21] The aim was to inhibit fruit softening while allowing other activities of fruit ripening such as colour and flavour development to proceed. This did not occur, and it was concluded that the decrease in polymerisation of pectin caused by polygalacturonase activity was not the cause of fruit softening. Nevertheless, these experiments describe a principle much more valuable than hard tomatoes, namely the discovery of new facts about a well studied phenomenon by altering a single component. This approach could be applied to many different aspects of plant metabolism in very productive ways.

FUTURE PROSPECTS

All of the work described has used single genes in improving agricultural traits. With an improved knowledge of metabolic pathways and plant development it should be possible to manipulate pathways for example, or redirect plant development along more desirable lines. Where information on the contribution of particular enzymes to metabolic pathways is known, then antisense methods can be used to alter the flux through the pathway, or alter the type of end product. The knowledge required of plant metabolism and development for such manipulations to be carried out is beyond our present understanding, therefore much effort needs to be put into studies of a fundamental nature. The power of genetics is being brought to bear on old problems of plant development, for example, responses to growth regulators, and this should provide new ways of understanding plant development. In addition, the concentration of effort in a few tractable plants such as *Arabidopsis* and tomato should speed up progress. A final impediment to progress is the current inability to transform several major crop plants such as wheat and barley, but rapid progress is being made towards this goal.

REFERENCES

1. Klein, T. M., Wolf, E. D., Wu, R. & Sanford, J. C., High velocity microprojectiles for delivering nucleic acids into living cells. *Nature,* **327** (1987) 70–3.
2. Weising, K., Schell, J. & Kahl, G., Foreign genes in plants: structure, expression and applications. *Ann. Rev. Genet.,* **22** (1988) 421–77.
3. Herrera-Estrella, L., van den Broeck, G., Maenhaut, R., van Montagu, M. & Schell, J., Light inducible and chloroplast-associated of a chimaeric gene introduced into tobacco using Ti plasmid vectors. *Nature,* **330** (1984) 160–3.
4. Jefferson, R. A., Kavanagh, T. & Bevan, M. W., GUS fusions: β-glucuronidase as a sensitive and versatile gene fusion marker in higher plants. *EMBO J.,* **6** (1987) 3901–7.
5. Kay, R., Chen, A., Daly, M. & McPherson, J., Duplication of CaMV 35S

promoter creates a strong enhancer for plant genes. *Science,* **236** (1987) 1299–302.

6. Fromm, M., Taylor, L. P. & Walbot, V., Stable transformation of maize after gene transfer by elecroporation. *Nature,* **319** (1986) 791–3.

7. Gallie, D. R., Sleat, D. E., Watts, J., Turner, P. C. & Wilson, T. M. V., A comparison of eukaryotic viral 5′ leader sequences as enhancers of translation in vivo. *Nucl. Acids Res.,* **15** (1987) 8693–711.

8. Beachy, R. N., Chen, Z. I., Horsch, R., Rogers, S. G., Hoffmann, N. J. & Fraley, R. T., Accumulation and assembly of soybean β-conglycinin in seeds of transformed petunia. *EMBO J.,* **4** (1985) 3047–53.

9. Iturriaga, G. & Bevan, M. W., Endoplasmic reticulum targeting and glycosylation of hybrid proteins in transgenic tobacco. *The Plant Cell,* **1** (1989) 381–90.

10. Shah, D. M., Horsch, R. B., Klee, H. J., Kishore, G. M., Winter, J. A., Hoffmann, N. J. & Fraley, R. T., Engineering herbicide tolerance in plants. *Science,* **233** (1986) 478–81.

11. Comai, L., Facciotti, D., Hiatt, W. R., Thompson, G., Rose, R. E. & Stacker, D. M., Expression of a mutant *aroA* gene from *Salmonella* in plants confers resistance to glyphosate. *Nature,* **317** (1985) 741–4.

12. Lee, K. Y., Townsend, J., Tepperman, J., Black, M., Chui, C. H., Mazur, B., Dunsmuir, P. & Bedbrook, J., The molecular basis of sulphonylurea herbicide resistance in tobacco. *EMBO J.,* **7** (1988) 1241–8.

13. Haughn, G. W. & Sommerville, C., Mutants of *Arabidopsis* resistant to sulphonylurea herbicides. *Molec. Gen. Genet.,* **204** (1986) 430–4.

14. Stalker, D. M., McBride, K. E. & Malgi, C. D., Herbicide resistance in transgenic plants expressing a bacterial detoxification gene. *Science,* **242** (1988) 419–23.

15. De Block, M., Botterman, J., Vanderweile, M., Dockx, J., Thoen, C., Gossele, V., Rao Movva, N., Thompson, C., van Montagu, M. & Leemans, J., Engineering herbicide resistance in plants by expressing a detoxifying enzyme. *EMBO J.,* **6** (1987) 2513–18.

16. Vaeck, M., Reynaertz, A., Hofte, H., Jansens, S., De Beuckeleer, M., Dean, C., Zabean, M., van Montagu, M. & Leemans, J., Transgenic plants protected from insect attack. *Nature,* **328** (1987) 33–7.

17. Hilder, V. A., Gatehouse, A. M. R., Sheerman, S. E., Barker, R. F. & Boulter, D., A novel mechanism of insect resistance engineered into tobacco. *Nature,* **330** (1987) 160–3.

18. Powell-Abel, P., Nelson, R. S., De, B., Hoffmann, N., Rogers, S. G., Fraley, R. T. & Beachy, R. N., Delay of disease development in transgenic plants that express the TMV coat protein gene. *Science,* **232** (1986) 738–43.

19. Hemenway, C., Fang, R. X., Kaniewski, W., Chua, N.-H. & Tumer, N. E., Analysis of the mechanism of protection in transgenic plants expressing the PVX coat protein gene. *EMBO J.,* **7** (1988) 1273–80.

20. Baulcombe, D. C., Saunders, G. R., Bevan, M. W., Mayo, M. R. & Harrison, B. D., Expression of a biologically active viral satellite RNA from the nuclear genome of higher plants. *Nature,* **321** (1986) 446–9.

21. Smith, C. J. S., Watson, C. F., Ray, J., Bird, C., Morris, P. C., Schuch, W. & Grierson, D., Antisense RNA inhibition of polygalacturonase gene expression in transgenic tomatoes. *Nature,* **334** (1988) 724–6.

Chapter 13

LARGE SCALE CULTURE OF MAMMALIAN CELLS FOR PRODUCTION OF THERAPEUTIC PROTEINS

J. R. BIRCH

Celltech Ltd, Slough, Berkshire, UK

INTRODUCTION

Animal cell technology has been playing an important role in the production of therapeutic products for more than 30 years. Following the introduction of polio vaccine in 1954 a number of other vaccines have been produced in animal cell culture, including measles, mumps and rubella. The technology has also been applied to the production of veterinary products such as foot-and-mouth disease virus vaccine. This particular vaccine, made in baby hamster kidney (BHK) cells, still represents one of the largest animal cell processes with millions of litres of culture fluid a year being processed.[1] During the 1970s animal cell culture was developed for the manufacture of some natural cell products which are now used clinically; such as urokinase and interferon β. The early human therapeutic products were manufactured in 'normal' cells, typically human diploid fibroblasts. These cell types have limited utility for applications requiring very large scale manufacture because they have a finite lifespan, are anchorage dependent and tend to require complex (and therefore expensive) culture media. In recent years attention has turned to the use of immortal cell lines such as myelomas and Chinese Hamster Ovary (CHO) cells which have an infinite lifespan, usually grow in suspension and can be cultured in less complex media than 'normal' cells. Such cell lines lend themselves much more readily to large scale manufacturing. The first example of the use of such a process for the production of a human therapeutic protein was in the manufacture of human lymphoblastoid interferon. This is made from a permanent human lymphoblastoid cell line (Namalva) grown in 8000 litre reactors.[2] One particular issue which has to be addressed in using permanent cell lines is that of product safety. In developing processes which use this type of cell particular attention has to be paid to the elimination of risk associated with potential viral and DNA contaminant hazards. This has in itself been an area of major technological advance in the last decade. The use of permanent cell lines contributed greatly to our ability to make interesting proteins on a large scale but was not the complete solution. For most proteins of interest, the cells producing the product either could not be cultured or, if they could, their productivity was very low. This situation changed dramatically with the advent of the technologies of cell fusion and recombinant DNA. For the first time a wide range of mammalian proteins with commercial potential could be made in cell culture using optimised expression systems[3] to generate large quantities of product. Table 1 compares the productivity of genetically manipulated cells making tissue plasminogen activator (t-PA) and interferon β with cells which naturally produce these proteins. In both cases the recombinant DNA approach gives an increase in productivity of several hundred fold. To put this in perspective, a single therapeutic dose of t-PA (*c*. 50 mg) would require many hundreds of litres of culture fluid if made in melanoma cells which naturally produce the protein.

In some cases recombinant therapeutic proteins (for example, growth hormone and interferons) have been produced successfully in micro-organisms. Despite the apparent economic advantages of using microbial systems there are an increasing number of situations where mammalian cell culture is preferred. This is because proteins derived from mammalian cells are generally correctly folded, have the correct post

Table 1
Protein Production in Naturally Producing and Genetically Engineered Cell Lines

Product	Cell type	Productivity in culture	Reference
(a) Interferon	Normal human fibroblasts	4×10^7 U/litre	21
	Engineered CHO cells	10^{10} U/litre	22
(b) Tissue plasminogen activator	Human melanoma cells	0·2 mg/litre	23
	Engineered myeloma cells	55 mg/litre	7

translational modifications (especially glycosylation) and are usually secreted in a functional form, in contrast to proteins made in bacteria. The advantages of using mammalian cells are particularly apparent for large complex proteins (such as t-PA) and for proteins where modification such as glycosylation is important.

Examples of products made in animal cells using recombinant DNA or hybridoma technology are shown in Table 2. Several of these products (OKT3, t-PA, erythropoietin) are already approved in various countries. New therapeutic products of animal cells are expected to make a significant commercial impact with sales of many hundreds of millions of dollars a year predicted.[4,5]

The use of recombinant DNA techniques now allows us to make not just natural proteins but completely novel molecules. For example, it is now possible to make chimeric mouse–human antibodies combining a mouse variable region with a human constant region (reviewed by Williams[6]). These molecules are expected to be less immunogenic than mouse antibodies when used therapeutically. An example of the production of such an antibody will be given later.

REACTORS FOR ANIMAL CELL CULTURE

The development of novel products from mammalian cell culture has brought with it the need to develop efficient, economic manufacturing processes. At present there are few applications requiring more than a kilogram or two of product per year, although it is likely that tens and even hundreds of kilograms of particular proteins will be required. In considering the types of reactor which we can use for mammalian cell culture we have to distinguish between cells which grow in suspension

Table 2
Mammalian Cell Products

Product	Clinical application	Cell type	Reference
Monoclonal antibody (OKT3)	Renal transplants	Murine Hybridoma	24
Tissue plasminogen activator (t-PA)	Coronary thrombosis	CHO	25, 26
Erythropoietin	Anaemia	CHO	27
Factor VIII	Haemophilia	Hamster kidney line	28, 29
Factor IX	Haemophilia	CHO	30
Hepatitis B surface antigen	Hepatitis B virus infection	CHO	31

(the majority of immortal cell lines) and those which are anchorage dependent (the majority of 'normal' cell lines). Examples of cells which grow in suspension include hybridomas and Chinese hamster ovary (CHO) cells (the cell line most commonly used to express recombinant proteins). Examples of anchorage dependent cells include human diploid fibroblasts used to produce virus vaccines. In some instances (for example CHO cells) cells can be grown either in suspension or attached to a surface.

Suspension Culture

In principle cells which grow in suspension can be grown in the same type of reactors used for microbial cultivation. Most of the existing and planned capacity for large scale mammalian cell culture is based on aerated and agitated bioreactors. These are based mainly on stirred tank and airlift designs. Suspension culture technology has several advantages — simplicity of scale-up and operation and good mixing which permits precise monitoring and control resulting in ease of optimisation and consistent production.

Stirred tank vessels ranging in working volume from 3000 to 10 000 litres are used for the production of foot-and-mouth disease virus vaccine,[1] lymphoblastoid interferon[2] and recombinant proteins. Airlift fermenters up to 2000 litre scale are used for the manufacture of monoclonal antibodies and recombinant proteins.[7] We have chosen to use airlift reactors for our large scale processes at Celltech. They are simply constructed as they lack mechanical agitation and are straightforward to scale up. These are important considerations if we are to benefit from economies of scale in capital and running costs.

In addition airlift fermenters are particularly efficient with respect to providing oxygen for cell growth, even at large scale.[8] Figure 1 shows a 2000 litre airlift fermenter and associated equipment. The airlift principle (reviewed by Onken and Weiland[9]) has been applied to microbial culture for many years and was first used for mammalian cell culture by Katinger *et al*.[10] Gas mixtures are introduced into the reactor from a sparge tube at the base of a central draught tube. This sets the culture into circulation providing mixing and also supplying oxygen to the culture. An outline of our 2000 litre production process is shown in Fig. 2. An inoculum culture grown in roller bottles is transferred aseptically to a 20 litre inoculum vessel. After an appropriate period of growth this inoculum is transferred to a 200 litre fermenter which in turn is used to inoculate the 2000 litre production vessel. The fermenters are stainless steel pressure vessels which, with their associated pipework, can

Fig. 1. 2000 litre airlift fermenter (courtesy of Celltech Ltd).

be sterilised by steam *in situ*. Culture medium is sterilised by filtration into the reactors and air is similarly sterilised by filtration. During the fermentation, pH, dissolved oxygen and temperature are monitored and controlled. A computer is used to provide a high degree of automation. The computer controls all valve and pump activation sequences during cleaning, sterilisation, process control and harvesting; substantially reducing the labour required to operate the plant. Ancillary equipment includes a continuous centrifuge to separate cells and debris from

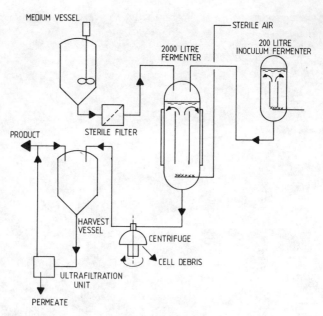

Fig. 2. 2000 litre airlift culture system. Reproduced with permission from Ref. 7. © 1988
Bio/Technology.

culture supernatant containing product; and an ultrafiltration system to concentrate product prior to purification. The quality of services is particularly important, especially when therapeutic proteins are being made and, for example, water is used for injection (WFI) for medium preparation.

Figure 3 shows an example of recombinant product formation in an airlift fermenter. In this example the product is a mouse–human chimeric B72.3 antibody.[11]

The B72.3 antibody recognises an epitope on a human tumour associated antigen and has been used for in-vivo detection of colon carcinoma. The recombinant protein is expressed in CHO cells grown in fed batch culture. During the fermentation concentrated solutions of key nutrients are fed to the culture to increase productivity of the culture. It will be seen that the total fermentation time was 8 days at which point the product had accumulated to a concentration of 41 mg/litre. This is a typical concentration for a recombinant protein produced in cell culture.

Airlift reactors can be operated either batchwise or continuously. Batch processes are most commonly used in industry because they are the simplest (and hence most reliable) and because alternative processes in general offer no economic advantages at present.

A plethora of alternative systems for cell culture have been developed

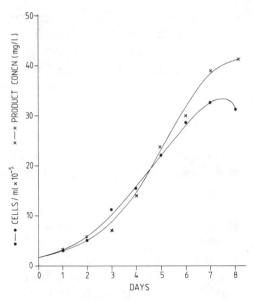

Fig. 3. Production of recombinant B72.3 chimaeric antibody in CHO cells growing in 100 litre airlift fermenter.

on a small scale. These are based, for example, on cell encapsulation, and entrapment in hollow fibres (see Refs 12 and 13 for reviews). There is at present no evidence that these systems offer any economic advantages over the simple systems already used in industry and it remains to be seen whether they will find wider application for large scale manufacture.

Anchorage Dependent Cells

In the case of anchorage dependent cells the most successful large scale technology to date has used the microcarrier technique developed by van Wezel.[14, 15] In this technique cells are grown on small beads which in turn are held in suspension in conventional fermenters, allowing one to benefit from the advantages described above for such reactors. The technology is being developed for a range of products. To date the largest scale applications (1000–4000 litres) have been for the production of interferon and polio vaccine.[16] When microcarrier culture is used for recombinant protein production there is an advantage in operating a continuous perfused process to maximise use of the microcarrier beads which are very expensive.

A microcarrier process for t-PA production from mouse C127 cells transfected with a bovine papilloma virus DNA based vector for expression of the human t-PA gene has been described.[17] The cells were cultured in a continuously perfused, spin filter stirrer reactor. High levels of tPA production (>50 mg/litre at a perfusion rate of 0·5 fermenter volumes per day) were achieved in cultures lasting up to 1000 h.

Apart from microcarrier culture, the most commonly used technique, particularly for virus vaccine production, is roller bottle culture (for example see Ref. 18).

Cell Culture Media

Animal cell culture media typically contain around fifty defined components and are frequently supplemented with animal serum. Serum is ill defined and is the most expensive component of the medium. In addition the presence of serum protein complicates the purification of protein products. Hence the culture medium is a major factor influencing cost and quality of product at large scale. In the last few years great progress has been made in elucidating the role of serum and many cell types can now be grown in the absence of serum.[19] The author's company routinely grows hybridoma and CHO cells in serum free medium for production. Recently a medium for hybridoma culture has been developed in which several of the proteins traditionally used in serum free medium have been reduced or eliminated. The protein concentration in this medium is 1 mg/litre compared with 1 g/litre in the original serum free medium. Since monoclonal antibodies typically accumulate to concentrations of 100–200 mg/litre, the protein in the culture medium now represents a small proportion of the total protein to be processed.

PRODUCT RECOVERY AND QUALITY ISSUES

In general protein products of mammalian cells are secreted into low protein or protein free media. This is a significant benefit in designing efficient recovery processes. Since the product is usually in very dilute solution at the end of a fermentation process, the first stage of recovery is usually a concentration step, for example, using ultrafiltration. This is typically followed by a series of high resolution chromatography steps leading to a high level of biochemical purity. A wide range of physicochemical and immunological techniques are used to quantify the level of purity.

In addition to biochemical purity it is also necessary to ensure that biologically active agents such as viruses are not transferred to the product. The use of immortal cell lines raises particular regulatory issues because these cells may give rise to tumours in animal models. Hence particular attention has to be paid to establishing freedom of the product from viruses and DNA which might have oncogenic potential. In addition to testing the product and the highly characterised cell banks used for production processes are validated for their ability to remove DNA and viruses. The *Guidelines on the Production and Quality Control of Monoclonal Antibodies of Murine Origin Intended for Use in Man* produced by the Committee for Proprietary Medicinal Products[20] gives further insight into the ways in which these regulatory issues are addressed in the particular case of antibodies.

REFERENCES

1. Radlett, P. J., Pay, T. W. F. & Garland, A. H. M., The use of BHK suspension cells for the commercial production of foot-and-mouth disease vaccines over a twenty year period. *Dev. Biol. Stand.,* **60** (1985) 163–70.
2. Pullen, K. F., Johnson, M. D., Phillips, A. W., Ball, G. D. & Finter, N. B., Very large scale suspension cultures of mammalian cells. *Dev. Biol. Stand.,* **60** (1985) 175–7.
3. Bebbington, C. & Hentschel, C., The expression of recombinant DNA products in mammalian cells. *Trends Biotechnol.,* **3** (1985) 314–17.
4. Klausner, A., Handicapping pharmaceuticals prospects. *Bio/Technology,* **5** (1987) 116.
5. Ratafia, M., Mammalian cell culture: worldwide activities and markets. *Bio/Technology,* **5** (1987) 692–4.
6. Williams, G., Novel antibody reagents: production and potential. *Trends Biotechnol.,* **6** (1988) 36–42.
7. Rhodes, M., & Birch, J. R., Large-scale production of proteins from mammalian cells. *Bio/Technology,* **6** (1988) 518–23.
8. Birch, J. R., Lambert, K., Thompson, P. W., Kenney, A. C. & Wood, L. A., Antibody production with airlift fermenters. In *Large Scale Cell Culture Technology,* ed. K. Lydersen, Hanser Publishers, Munich, Vienna and New York, 1987, pp. 1–20.
9. Onken, U. & Weiland, P., Airlift fermenters: construction, behaviour and uses. *Adv. Biotechnol. Proc.,* **1** (1983) 67–95.
10. Katinger, H. W. D., Scheirer, W. & Kromer, E., Bubble column reactor for mass propagation of animal cells in suspension culture. *Ger. Chem. Engng* (Engl. Transl.), **2** (1979) 31–8.
11. Whittle, N., Adair, J., Lloyd, C., Jenkins, L., Devine, J., Schlom, J., Raubitschek, A., Colcher, D. & Bodmer, M., Expression in COS cells of a mouse–human chimeric B72.3 antibody. *Prot. Engng,* **1** (1987) 499–505.
12. Lydersen, K. (ed.), *Large Scale Cell Culture Technology.* Hanser Publishers, Munich, 1987.
13. Feder, J. & Tolbert, W. R., *Large Scale Mammalian Cell Culture.* Academic Press, New York, 1985.
14. van Wezel, A. L., Growth of cell strains and primary cells on microcarriers in homogeneous culture. *Nature, Lond.,* **216** (1967) 64–5.
15. van Wezel, A. L., Monolayer growth systems: homogeneous unit processes.

In *Animal Cell Biotechnology*. Vol. 1, ed. R. E. Spier & J. B. Griffiths. Academic Press, New York, 1985, pp. 266–82.

16. Montagnon, B. J., Vincent-Galquet, J. C. & Janget, B., Thousand litre scale microcarrier culture of Vero cells for killed poliovirus vaccine. Promising results. *Dev. Biol. Stand.,* **47** (1983) 55–64.

17. Birch, J. R. &. Rhodes, P. M., Large-scale production of proteins from mammalian cells. *Bio/Technology,* **6** (1988) 518–23.

18. Panina, G. F., Monolayer growth systems: Multiple processes. In *Animal Cell Biotechnology,* Vol. 1, ed. R. E. Spier & J. B. Griffiths. Academic Press, New York, 1985, pp. 211–42.

19. Lambert, K. J. & Birch, J. R., Cell growth media. In *Animal Cell Biotechnology,* Vol. 1, ed. R. E. Spier & J. B. Griffiths. Academic Press, New York, pp. 86–122.

20. Committee for Proprietary Medicinal Products Ad Hoc Working Party on Biotechnology/Pharmacy, Commission of the European Communities notes to applicants for marketing authorisations. *Guidelines on the Production and Quality Control of Monoclonal Antibodies of Murine Origin Intended for Use in Man. Trends Biotechnol.,* **6** (1988) G5–G8.

21. Mizrahi, A., Production of human interferons — an overview. *Proc. Biochem.* (August 1983) 9–12.

22. McCormick, F., Trahey, M., Innis, M., Dieckmann, B. & Ringold, G., Inducible expression of amplified human beta interferon genes in CHO cells. *Molec. Cell. Biol.,* **4** (1984) 166–72.

23. Kluft, C., van Wezel, A. L., van der Velden, C. A. M., Emeis, J. J., Verheijen, J. H. & Wijngaards, J. K., Large scale production of extrinsic (tissue-type) plasminogen activator from human melanoma cells. *Adv. Biotechnol. Proc.,* **2** (1983) 97–110.

24. Goldstein, G. & Ortho Multicenter Study Group, A randomised clinical trial of OKT3 monoclonal antibody for acute rejection of cadaveric renal transplants. *New Engl. J. Med.,* **313** (1985) 337–42.

25. Lubiniecki, A., Arathoon, R., Polastri, G., Thomas, J., Wiebe, M., Garnick, R., Jones, A., Van Reis, R. & Builder, S., Selected strategies for manufacture and control of recombinant tissue plasminogen activator prepared from cell cultures. *Proc. Eur. Soc. Animal Cell Technol.,* 1988, p. 100.

26. Birnbaum, S., Nilsson, K. & Mosbach, K., Production of human immune-interferon and human tissue-type plasminogen activator by microcarrier culture of recombinant chinese hamster ovary cells. *Proc. 4th European Congress on Biotechnology,* Vol 3, ed. O. M. Neijssel, R. R. van der Meer & K. Ch. A. M. Luyben. Elsevier, Amsterdam, 1987, p. 585.

27. Davis, J. M., Arakawa, T., Strickland, T. W. & Yphantis, D. A., Characterisation of recombinant human erythropoietin produced in Chinese Hamster Ovary cells. *Biochemistry,* **26** (1987) 2633–8.

28. Wood, W. I., Capon, D. J., Simonsen, C. C., Eaton, D. L., Gitschier, J., Keyt, B., Seeburg, P. H., Smith, D. H., Hollingshead, P., Wion, K. L., Delwart, E., Tuddenham, E. G. C., Vehar, G. A. & Lawn, R. M., Expression of active human factor VIII from recombinant DNA clones. *Nature, Lond.,* **312** (1984) 330–7.

29. Toole, J. J., Knopf, J. L., Wozney, J. M., Sultzman, L. A., Buecker, J. L., Pitmann, D. D., Kaufman, R. J., Brown, E., Shoemaker, C., Orr, E. C., Amphlett, G. W., Foster, W. B., Coe, M. L., Knutson, G. J., Fass, D. N. &

Hewick, R. M., Molecular cloning of a cDNA encoding human anti-haemophilic factor. *Nature,* **312** (1984) 342–7.

30. Kaufman, R. J., Wasley, L. C., Furie, B. C. & Shoemaker, C. B., Expression, purification and characterization of recombinant γ carboxylated Factor IX synthesized in Chinese Hamster Ovary cells. *J. Biol. Chem.,* **261** (1986) 9622–8.

31. Michel, M.-L., Sobczak, E., Malpiece, Y., Tiollais, P. & Streek, R. E., Expression of amplified Hepatitis B virus surface antigen genes in Chinese Hamster Ovary cells. *Bio/Technology,* **3** (1985) 561–6.

Chapter 14

THE ROLE OF MOLECULAR BIOLOGY IN DRUG DISCOVERY AND DESIGN

DOLAN B. PRITCHETT & PETER H. SEEBURG

Department of Molecular Neuroendocrinology, Zentrum für Molekulare Biologie (ZMBH), Heidelberg, FRG

INTRODUCTION

To be able to design a chemical compound that would interact with a characterized receptor on a cell in a specific and predictable way has been the stated goal of pharmacology from very early on. Such rational drug design would in fact be a great achievement. However, we can expect that many new specific drugs will also be discovered by testing of compounds as currently practised. What we propose to do in this chapter is to examine how recent developments in molecular biology have provided several important techniques which can be used to study the interactions of drugs with specific receptor subtypes and how these same systems can be used to study the structure of specific receptors. Hopefully this will lead to the goal of rational drug design as well as provide new and more specific screening procedures for new drugs.

In order to move towards the goal of rational drug design it is necessary to know as much as possible about the structure of the drug itself and the structure of the receptor. To know the active structure of even a simple drug is certainly a complex task but it is simple relative to the task of knowing the structure of a binding site on the receptor, especially considering the fact that most receptors are membrane proteins and therefore the theoretically straightforward approach of X-ray crystallography may be practically very difficult.

RECEPTOR FAMILIES

A common first step in the analysis of the structure of a receptor is to ascertain the primary sequence. Here molecular biology has produced the expected result of predicting the complete amino acid sequence from clones isolated by using DNA probes coding for small peptides (receptor fragments) which have been purified and sequenced. This has greatly facilitated the task of determining the primary structure of several receptors. More importantly though, molecular biologists have been able to determine the primary structure of receptors for which no amino acid sequence is available. This has been done by taking advantage of the fact that most receptors discovered to date are members of gene families.[1-3] In other words the receptor proteins share certain conserved amino acid sequences. Using these conserved sequences molecular biologists have been able to find other members of the gene family. For example the cDNAs encoding the dopamine D2 receptor and 5HT1A (serotonin) receptor were recently cloned by using radiolabeled β-adrenergic receptor cDNA as a hybridization probe.[4,5] The short regions of similarity between these genes were enough to identify new members of the receptor gene family and subsequent testing characterized the gene products as the dopamine D2 and the 5HT1A receptors. A similar strategy has been used in our laboratory to clone the serotonin $5HT_2$ receptor[6] and subunits of the $GABA_A$ receptor.[7-9] Other examples are indicated in Table 1 which show the receptor gene families and their members. In all cases it is noteworthy that pharmacologically so very distinct receptors were identified on the basis of their structural similarities. This reflects the fact that neurotransmitter receptors fall into two main structural classes. One class consists of single glycoproteins containing seven transmembrane regions and the other class of ion channel complexes consisting of several subunits. These classes differ structurally from the class of growth factor receptors and that of voltage-clamped ion channels.

The simplification that comes as a result of being able to group these receptors into families based on common structural motifs ironically has produced added complexity. These structural similarities have been used to isolate receptor subtypes which were not known from the pharmacological data. The muscarinic receptor is a good example of this phenomenon.[10-13] Two muscarinic subtypes were known to exist but recent work in several laboratories including our own has demonstrated three additional receptor subtypes. The $GABA_A$ receptor family, which is a member of a different class, contains more than fourteen genes[14] while only two well-characterized subtypes had been described. Molecular biology has produced 'good news' and 'bad news' if you will. The good news is that the diversity of receptors can be simplified into four large gene families. Thus, what is found for one receptor will likely be very useful in understanding other members of its family. The bad news is

Table 1

Members of Various Receptor Genes Grouped into Gene Families

G-Protein coupled seven transmembrane domains	*Ligand-gated ion channels four transmembrane domains*	*Tyrosine kinase single transmembrane domain*
Dictyostelium cAMP receptor[17]	Nicotinic acetylcholine[26]	Insulin (reviewed in Ref. 3)
Angiotensin[18]	GABA$_A$[1,7-9]	HER2/NEU
Yeast mating factor[19]	Glycine[27]	Epidermal growth factor
Substance K[20]	Neuronal nicotinic acetylcholine[28]	Insulin-like growth factor I
Rhodopsin[21]		Platelet derived growth factor I
Muscarinic M1, M2, M3, M4, M5[10-13]		c-KIT oncogene
Dopamine D2[4]		Macrophage growth factor CSF
Adrenergic α1, β1, β2[22-24]		
Serotonin 5HT-2, 5HT-1c, 5HT-1a[5,6,25]		

The members of each family are grouped under a heading that indicates their method of signalling and their characteristic structural motif. It is important to note that these characteristics have not been demonstrated experimentally in all cases but are predicted by analogy. Superior numbers specify relevant references.

that the diversity of receptors may be substantially larger than anticipated. From a pharmacological point of view, the bad news may be good news as well. If drugs could be designed which interact specifically with one of the new subtypes, these may be more useful drugs.

RECOMBINANT RECEPTOR ANALYSIS

This returns us to the problem of designing or discovering drugs specific for these receptor subtypes. Here again molecular biology could play a major role, though that role remains untested to date. To screen drugs for interactions with specific receptors or to determine the structure of receptors it is important to isolate the receptor of interest from other receptor subtypes and to reconstitute the receptor into a functional environment. The way this has been done is to express the cloned cDNA in one of several systems. This chapter will concentrate on expression of cloned receptors in mammalian cells by transient DNA-mediated transfection.[15] It will also deal almost exclusively with members of the ligand-gated ion channel family and the G protein-coupled receptor family, although the conclusions drawn can be applied in most cases to the other receptor families.

There are several advantages of this transient expression system. First, it is possible to characterize the receptors in several cell types before choosing a particular cell for stable expression. The wide variety of cell types almost assures the possibility of finding a cell type without competing for endogenous receptors. Second, it is easy to obtain a sufficient amount of membranes to look at the binding characteristics of any radioactive ligand. Third, the same cells can also be characterized electrophysiologically using the sensitive patch-clamp technique and drugs can be rapidly applied over the whole cell surface. These advantages are amplified by the relative simplicity of tissue culture and DNA transfection.

Outlined in Fig. 1 is the process of cell transfection and selection for stable cell lines. Whether one chooses to do experiments in transiently transfected cell lines depends on the experiment. In general, preliminary experiments are carried out by transient transfection and when it is clear which cell and receptor combinations work well the stable cell lines can be constructed. In addition to the time (1 month) it takes to make stable cell lines, there has been one major problem reported: the cell lines do not continue to produce high numbers of receptors as the lines are cultured. Apparently, expressing certain receptors, specifically ligand-gated ion channels under normal culture conditions place expressing

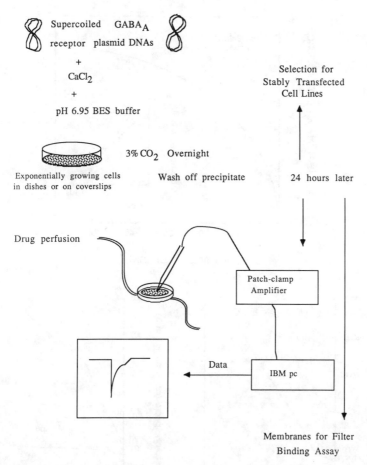

Supercoiled GABA$_A$ receptor plasmid DNAs

+

CaCl$_2$

+

pH 6.95 BES buffer

3% CO$_2$ Overnight

Exponentially growing cells in dishes or on coverslips

Wash off precipitate

Selection for Stably Transfected Cell Lines

24 hours later

Drug perfusion

Patch-clamp Amplifier

Data

IBM pc

Membranes for Filter Binding Assay

Fig. 1. Schematic representation of the expression and characterization of receptors in DNA transfected mammalian cells. A detailed description of the steps involved can be found in Refs 8 and 15.

cells at a growth disadvantage. This may explain why there are few or no naturally occurring cell lines which express high numbers of gated channels. However, if large amounts of cells are needed, a stable cell line is obviously worth the investment of resources. The transient system has several disadvantages. One of these is the constant need for supercoiled (expression vector) plasmid DNA. Since as much as 20 μg of DNA are used per 10 cm plate the constant supply of transiently transfected cells can be very time consuming. Also, to use the cells in electrophysiology, a large proportion of the cells should be expressing the receptor. Such transfection efficiencies may not be possible with all cell lines.

In Fig. 2 some of the data that can be obtained using the transient cell transfection system is presented. Data are shown for both a member of

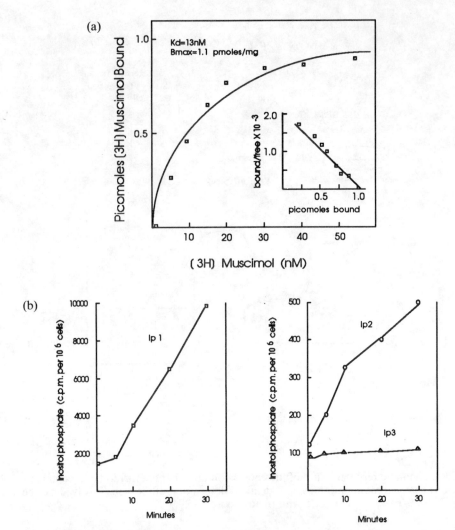

Fig. 2. Data obtained from experiments using transiently transfected mammalian cells. (a) The saturation binding experiments and Scatchard plot of data (insert) using ^{3}H-muscimol and cell membranes from GABA receptor cDNA transfected cells.[15] (b) The serotonin stimulated production of inositol phosphates in cells transfected with the 5HT$_2$ receptor cDNA. The level of radioactivity in each of the fractions of an ion exchange column is plotted versus time.[16]

the G-protein coupled family, 5HT$_2$, and the ligand-gated ion channel family, the GABA$_A$ receptor. These examples show how this system can be used to study not only the binding of drugs to receptors of various types but also the functional coupling of this binding to obtain a measurable effect. It is important when screening for new drugs as well as testing existing ones to know that the drug not only binds but whether it is an agonist or an antagonist. Again, by choosing the cell type carefully many different aspects of the receptor function can be studied and the method of drug modulation of each of these effects can be studied.

As can be seen from Fig. 2, different receptors can be studied by cell expression. That this system can be used to examine receptor subtypes is shown in Fig. 3. It had been shown from in-vivo studies that not all $GABA_A$ receptors responded similarly to pentobarbital. Differences in the response to this drug were found only when measuring from certain brain regions but proof that there were subtypes of the $GABA_A$ receptor based on these measurements would have been tedious. However, when different clones of the $GABA_A$ receptor subunits were expressed in cells a consistent effect could be seen and this effect could be correlated with a certain receptor type containing the α_2 receptor subunit. These studies show that receptor subtypes that are thought to exist in-vivo can be faithfully reproduced by cells expressing transfected DNAs. The number of amino acids which differ between α_1 and α_2 subunits are quite small. Thus, the likelihood of discovering such receptor subtypes by other methods is low. Hence, the methods outlined in Figs 2 and 3 prove the usefulness of this system for studying the interaction of drugs and receptor subtypes. Screening of a large number of compounds for binding as well as for agonist or antagonist effects could be accomplished by a scale up of the techniques shown in Figs 2 and 3.

OUTLOOK

As was mentioned earlier, molecular biology will certainly play a further role in characterizing drug binding sites. The availability of cloned DNA

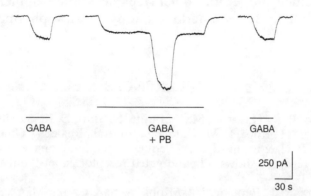

Fig. 3. Current traces showing the effect of pentobarbitol on cells transfected with the $\alpha 2$ subtype with the β and γ subunits of the $GABA_A$ receptor. Application of $50 \mu M$ pentobarbitol indicated by the bar causes an inward current only in the $\alpha 2$-transfected cells but not in the $\alpha 1$-transfected cells. The cell was voltage clamped at -60 mV in the whole cell configuration and drugs were applied for the times indicated by the bars.[15]

and the ability to express receptors at high levels (at the present time at least transiently) provides an experimental framework in which the structure of the receptor can be studied by making changes in the receptor through in-vitro mutagenesis. Then, the effect of the changes can be studied at the level of binding and coupling studies. This is a method that is only being applied to a few receptors so far, but as the sequence of more receptors become available it will provide more information on which to base mutational changes. Again, any information gained about the binding site of one receptor will have important implications for the characterization of other receptors in its family.

There is also reason to expect that molecular biological techniques will be applied in combination with traditional biochemical and biophysical techniques to provide large amounts of receptor protein necessary for structural studies. Since the abundance of the receptors naturally is often quite low, artificial systems such as expression in bacterial, mammalian or insect cells may be required to obtain sufficient amounts of receptor. These systems are beginning to be used for the study of receptors and will prove very useful in the future.

In conclusion, the authors feel that molecular biology has already made important contributions to the study of drug receptors but are confident that recent developments in the expression of these receptors will lead to many further contributions to the understanding of drug–receptor interactions. Specifically, it is predicted that transfected cell expression of panels of receptor subtypes will provide a new screening approach for clinically important specific drugs, and that these same systems will be used to study the structure of receptors. This, it is believed, will lead to the discovery and design of pharmaceuticals which display higher specificity, fewer unwanted side effects and will therefore prove more useful than those hitherto found by classical pharmacology.

REFERENCES

1. Schofield, P. R., Darlison, M. G., Fujita, N., Burt, D. R., Stephenson, F. A., Rodriguez, H., Rhee, L. M., Ramachandran, J., Reale, V., Glencorse, T. A., Seeburg, P. H. & Barnard, E. A., Sequence and functional expression of the GABA$_A$ receptor shows a ligand-gated receptor super-family. *Nature,* **328** (1987) 221-7.
2. Lefkowitz, R. J., Benovic, J. L., Kobilka, B. & Caron, M. G., β-adrenergic receptors and rhodopsin: shedding light on an old subject. *Trends Pharmacol. Sci.,* **7** (1986) 444-8.
3. Yarden, Y. & Ullrich, A., Growth factor receptor tyrosine kinases. *Ann. Rev. Biochem.,* **57** (1988) 443-78.

4. Bunzow, J. R., Van Tol, H. H. M., Grandy, D. K., Albert, P., Salon, J., Christie, M., Machida, C. A., Neve, K. A. & Civelli, O., Cloning and expression of a rat D2 dopamine receptor cDNA. *Nature,* **336** (1988) 783–7.
5. Fargin, A., Raymond, J. R., Lohse, M. J., Kobilka, B. K. & Caron, M. G., The genomic clone G-21 which resembles a β-adrenergic receptor sequence encodes the 5-HT$_{1A}$ receptor. *Nature,* **335** (1988) 358–60.
6. Pritchett, D. B., Bach, A. W. J., Wozny, M., Taleb, O., DalToso, R., Shih, J. C. & Seeburg, P. H., Structure and functional expression of cloned rat serotonin 5HT-2 receptor. *EMBO J.,* **7** (1988) 4135–40.
7. Levitan, E. S., Schofield, P. R., Burt, D. R., Rhee, L. M., Wisden, W., Köhler, M., Fujita, N., Rodriguez, H. F., Stephenson, A., Darlison, M. G., Barnard, E. A. & Seeburg, P. H., Structural and functional basis for GABA$_A$ receptor heterogeneity. *Nature,* **335** (1988) 76–9.
8. Pritchett, D. B., Sontheimer, H., Shivers, B. D., Ymer, S., Kettenmann, H., Schofield, P. R. & Seeburg, P. H., Importance of a novel GABA$_A$ receptor subunit for benzodiazepine pharmacology. *Nature,* **338** (1989) 582–5.
9. Ymer, S., Schofield, P. R., Draguhn, A., Werner, P., Köhler, M., and Seeburg, P. H., GABA$_A$ receptor β subunit heterogeneity: functional expression of cloned cDNAs. *EMBO J.,* **8** (1989) 1665–70.
10. Bonner, T. I., Buckley, N. J., Young, A. C. & Brann, M. R., Identification of a family of muscarinic acetylcholine receptor genes. *Science,* **237** (1987) 527–32.
11. Bonner, T. I., Young, A. C., Brann, M. R. & Buckley, N. J., Cloning and expression of the human and rat m5 muscarinic acetylcholine receptor genes. *Neuron,* **1** (1988) 403–10.
12. Braun, T., Schofield, P. R., Shivers, B. D., Pritchett, D. B. & Seeburg, P. H., A novel subtype of muscarinic receptor identified by homology screening. *Biochem. Biophys. Res. Comm.,* **149** (1987) 125–32.
13. Peralta, E. G., Ashkenazi, A., Winslow, J. W., Ramachandran, J. & Capon, D. J., Differential regulation of PI hydrolysis and denyl cyclase by muscarinic receptor subtypes. *Nature,* **334** (1988) 434–7.
14. Seeburg, P. H., unpublished observations.
15. Pritchett, D. B., Sontheimer, H., Gorman, C. M., Kettenmann, H., Seeburg, P. H. & Schofield, P. R., Transient expression shows ligand gating and allosteric potentiation of GABA$_A$ receptor subunits. *Science,* **242** (1988) 1306–8.
16. Pritchett, D. B., Bach, A., Taleb, O., DalToso, R. & Seeburg, P. H., Cloned serotonin 5HT-2 receptor: Structure and second messenger coupling. In *NATO ASI Series H32.* Springer-Verlag, Heidelberg, 1989, pp. 163–73.
17. Klein, P. S., Tzeli, J. S., Saxe, C. L., III, Kimmel, A. R., Johnson, R. L. & Devreotes, P. N., A chemoattractant receptor controls development in Dictyostelium discoideum. *Science,* **241** (1988) 1467–72.
18. Jackson, T. R., Blair, L. A. C., Marshall, J., Goedert, M. & Hanley, M. R., The mas oncogene encodes an angiotensin receptor. *Nature,* **335** (1988) 437–40.
19. Nakayama, Y., Miyajima & Arai, K., Nucleotide sequences of STE2 and STE3, cell type-specific sterile genes from Saccharomyces cerevisiae. *EMBO J.,* **4** (1985) 2643–8.
20. Masu, Y., Nakayama, K., Tamaki, H., Harada, Y., Kuno, M. & Nakanishi, S., cDNA cloning of bovine substance-K receptor through oocyte expression system. *Nature,* **329** (1987) 836–8.

21. Nathans, J. & Hogness, D. S., Isolation, sequence analysis and intron-exon arrangements of the gene encoding bovine rhodopsin. *Cell,* **34** (1983) 807–14.
22. Kobilka, B. K., Matsui, H., Kobilka, T. S., Yang-Feng, T. L., Francke, U., Caron, M. G., Lefkowitz, R. J. & Regan, J. W., Cloning, sequencing, and expression of the gene coding for the human platelet α_2-adrenergic receptor. *Science,* **238** (1987) 650–6.
23. Frielle, T., Collins, S., Kiefer, W. D., Caron, M. G., Lefkowitz, R. J. & Kobilka, B. K., Cloning of the cDNA for the human β_1-adrenergic receptor. *Proc. Nat. Acad. Sci., USA,* **84** (1987) 7920–4.
24. Dixon, R. A. F., Kobilka, B. K., Strader, D. J., Benovic, B. K., Dohlmann, H. G., Frielle, T., Bolanowski, M. A., Bennett, C. D., Rands, E., Diehl, R. E., Mumford, R. A., Slater, E. E., Sigal, I. S., Caron, M. G., Lefkowitz, R. J. & Strader, C. D., Cloning of the gene and cDNA for mammalian β-adrenergic receptor and homology with rhodopsin. *Nature,* **321** (1986) 75–9.
25. Julius, D., MacDermott, A. B., Axel, R. & Jessell, T. M., Molecular characterization of a functional cDNA encoding the serotonin 1c receptor. *Science,* **241** (1988) 558–64.
26. Noda, M., Furutani, Y., Takahashi, H., Toyosato, M., Tanabe, T., Shimizu, S., Kiyotani, S., Kayano, T., Hirose, T., Inayama, S. & Numa, S., Cloning and sequence analysis of calf cDNA and human genomic DNA encoding α-subunit precursor of muscle acetylcholine receptor. *Nature,* **305** (1983) 818–23.
27. Grenningloh, G., Rienitz, A., Schmitt, B., Methfessel, C., Zensen, M., Beyreuther, K., Gundelfinger, E. D. & Betz, H., The strychnine-binding subunit of the glycine receptor shows homology with nicotinic acetylcholine receptors. *Nature,* **328** (1987) 215–20.
28. Boulter, J., Evans, K., Goldman, D., Martin, G., Treco, D., Heinemann, S. & Patrick, J., Isolation of a cDNA clone for a possible neural nicotinic acetylcholine receptor α-subunit. *Nature,* **319** (1986) 368–74.

Chapter 15

HAEMOPOIETIC GROWTH FACTORS AS DRUGS

J. H. SCARFFE[a], W. P. STEWARD[a], N. G. TESTA[b] & T. M. DEXTER[b]

CRC Departments of [a]Medical Oncology, and [b]Experimental Hematology, Paterson Institute & Christie Hospital, Withington, Manchester, UK

INTRODUCTION

The process by which a small number of self renewing stems cells give rise to lineage committed progenitors that subsequently proliferate and differentiate to circulating mature blood cells is regulated by a family of glycoproteins known as colony stimulating factors (CSFs). Classification of these CSFs or growth factors, is based on the types of mature cells seen in colonies of bone marrow cells produced *in vitro* in response to these compounds. Thus, interleukin 3 (IL-3) stimulates the production of mature cells of most of the haemopoietic lineages including granulocytes, macrophages, eosinophils, megakaryocytes, erythroid cells and mast cells. Granulocyte-CSF (G-CSF) and macrophage-CSF (M-CSF) exhibit relative lineage restricted specificity and stimulate the production of granulocytes and macrophages respectively, while granulocyte-macrophage-CSF (GM-CSF) stimulates the production of both granulocytes and macrophages.

Recombinant DNA technology has recently enabled the production of sufficient quantities of the myeloid CSFs and erythropoietin (required for the production of red blood cells) to be evaluated in preclinical testing. Erythropoietin (EPO), GM-CSF and G-CSF have recently entered clinical studies in man. The potential use of haemopoietic growth factors are shown in Table 1. The preliminary results have been very exciting, and the use of these haemopoietic growth factors as drugs in the clinic may have considerable value in a variety of conditions.

The use of haemopoietic growth factors is a rapidly expanding field and many of the studies are preliminary. In this chapter, we review the results obtained to date and indicate lines of future development.

191

Table 1
Potential Clinical Applications of Myeloid CSFs

---

(1) Treatment of bone marrow failure:
    (a) idiopathic
    (b) neoplastic
    (c) iatrogenic
(2) Augment rate of recovery after bone marrow transplantation
(3) Reduce duration or degree of leucopenia following chemotherapy
(4) Increase granulocyte number and function (e.g. in patients with AIDS)
(5) Treatment of established bacterial and fungal infections
(6) Improve host defence against potential infection after major trauma (e.g. burns)
(7) Treatment of leukemia — alter the rates of self-reproduction and differentiation
(8) Treatment of myelodysplasia — increase normal differentiation and reduce blast population

---

ERYTHROPOIETIN

Patients with end-stage renal failure become profoundly anemic mainly as a result of the reduced production of erythropoietin. They may require intermittent red cell transfusions to maintain the hemoglobin at levels which allow them to carry out their usual daily activities. Unfortunately, repeated transfusions carry several risks including the transmission of infections, iron overload and the development of cytotoxic antibodies which can interfere with subsequent renal transplantation. The hope that the administration of erythropoietin would abrogate the need for transfusions in these patients made them an ideal group to use for the initial clinical trials of this growth factor. The first study was opened in December 1985.

Reports of the initial two phase I/II studies of the use of erythropoietin in a total of 46 patients with end stage renal failure demonstrated dose-dependent rises in hemoglobin and hematocrit levels.[1,2] These findings were confirmed by a subsequent larger ongoing study for which a preliminary report of the first 250 patients is available.[3] All but one of the patients treated to date has become transfusion independent and all have reported increased exercise tolerance and sense of wellbeing as the anemia was corrected. Some patients have received erythropoietin for periods in excess of 2 years and they have remained sensitive to the hormone throughout without developing any detectable antibodies.

Acute side effects attributable to erythropoietin have been minimal. Transient mild dyspnoea, arthralgia or myalgia, flushing sensation and bone pain have occasionally been reported but always disappear after the first 2 or 3 doses have been given. Approximately 50% of patients experience increases of diastolic blood pressure and, early in the trials, some experienced seizures. With the realization that early institution of effective control of hypertension is required, this complication is becoming infrequent.

Studies have been established to examine the administration of erythropoietin to patients with anemia associated with chronic diseases (such as rheumatoid arthritis). Preliminary communications suggest encouraging results in these conditions. Neoplasia is often associated with anemia (either resulting from the underlying malignancy or from its treatment) and studies are planned to assess the role of erythropoietin in patients with a variety of neoplasms. A second indication for the use of erythropoietin in this group of patients may arise from the observation that at high doses it stimulates megakaryopoiesis *in vitro* and *in vivo*.[4] The potential of this agent to alleviate the thrombocytopenia induced by chemotherapy and to increase the rate of recovery from bone marrow transplantation is obviously of great interest. To date, erythropoietin has been given by intermittent intravenous bolus injections (e.g. at a dose of 150 units/kg/week in three divided doses for patients with renal failure) and further investigations into alternative routes and schedules of administration (e.g. subcutaneous injections or continuous intravenous infusion) may reveal more effective dose/response relationships.

Granulocyte-Macrophage Colony-Stimulating Factor

Three formal phase I studies to determine the toxicity and biological effects of GM-CSF in patients with normal bone marrow reserves have been reported. Herrmann *et al.*,[5] using glycosylated GM-CSF, compared the results of intravenous bolus and continuous infusion administration at six different dose levels (30–1000 $\mu g/m^2$/day). At 100 $\mu g/m^2$/day there was only a modest (two-fold) increment of the total white blood cell count (WBC) following bolus administration but a 17-fold increase was seen during continuous infusion at the same dose. Side effects were reported as mild and included low-grade fever, bone discomfort, myalgia, dyspnoea, nausea and headache. Similar disparities in response to the two methods of administration were seen in patients with primary or secondary bone marrow failure given GM-CSF as daily bolus injections or continuous intravenous infusions. No increase in leucocyte counts occurred with bolus injections but up to 8·5-fold rises in granulocyte counts resulted from the continuous infusion.[6] Recent work from our own group has confirmed the superiority of continuous intravenous infusions of GM-CSF as compared with intermittent bolus injections — both in terms of hematological response (up to 10-fold increment of WBC at doses of 3 and 10 μg/kg/day) (Fig. 1) and lack of any toxicity.

A phase I study with 20 patients given nonglycosylated GM-CSF as daily half hour intravenous bolus injections at doses of 0·3–60 μg/kg/day has recently been completed.[7] Increases (up to four-fold rises) of total WBC, eosinophil and neutrophil polymorphs were seen at dose levels of 10 μg/kg and above. Patients developed a neutropenia immediately after the start of administration of GM-CSF and the count returned to baseline levels 4-h later (Fig. 2). The subsequent rise of the count over the 10 days of daily injections occurred in a triphasic fashion. There was an early increase, probably due to demargination of cells, a subseqeunt plateau phase and a final rapid rise after the eighth day resulting from the appearance of leucocytes produced by cell proliferation induced in the bone marrow by GM-CSF. Toxic side effects included transient pyrexias, bone pain and pruritus. Severe toxicity occurred in four of the five patients who received a dose of 60 μg/kg and included pericarditis and a 'capillary leak' syndrome. One patient died from a pulmonary embolus although this may have been tumor related. Of great interest was the observation of significant tumor regression in one patient with a pretreated metastatic soft tissue sarcoma — an observation which may have important clinical and scientific implications.

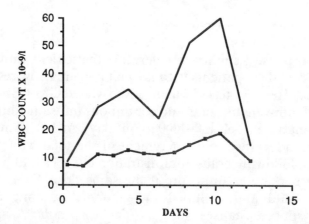

Fig. 1. Profile of total leucocyte count in a patient receiving GM-CSF given by daily intravenous half-hour bolus injections (—■—) at a dose of 10 μg/kg/day and, 4 months later, as a continuous infusion (———) at a dose of 3 μg/kg/day. The triphasic increase of peripheral leucocyte count seen after the administration of GM-CSF is illustrated with an initial early rise due to demargination of cells, a subsequent plateau phase and a final phase of rapid rise due to the appearance of leucocytes produced as a result of proliferation of bone marrow progenitor cells. The two curves show the superiority of continuous infusions over bolus injections with the former route producing a significantly higher rise of the white blood cell count even though the dose of GM-CSF was lower.

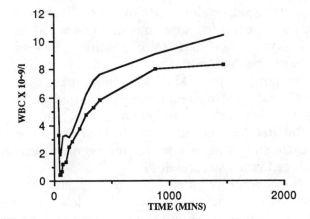

Fig. 2. Profile of change in total leucocyte (———) and neutrophil count (—■—) over 18 h after commencement of half hour intravenous infusion of GM-CSF. The counts fell rapidly within 5 min and subsequently returned to pretreatment levels 4 h later. Higher values are found subsequently.[7]

Responses to subcutaneous injections of non-glycosylated GM-CSF were examined in the third phase I study.[8] This route of administration appears to be more effective than intermittent bolus intravenous injections resulting in five-fold rises of total leucocyte counts at a dose of 15 µg/kg given daily for up to 10 days. Administration by this route was generally well tolerated apart from occasional skin rashes at injection sites.

These phase I studies confirm in-vitro observations that continuous exposure to CSFs is required to stimulate cell proliferation and functional activity. The half-life of intravenously injected GM-CSF is only 1–3 h[9] whereas subcutaneous injections, even at doses of 3 µg/kg, produce serum levels above 1 ng/ml (the concentration required to give >90% of maximal cell proliferation *in vitro*) for 10 or more hours.[8] Optimal administration for future clinical trials will, therefore, have to be via the subcutaneous route (perhaps twice daily) or by continuous intravenous infusion.

Phase I/II studies in humans have been established to determine the effects of GM-CSF on patients with bone marrow suppression resulting from a variety of causes (chemotherapy, myelodysplasia, aplastic anemia) and encouraging results have been obtained. Multilineage responses to continuous infusions of this growth factor (at doses of 15–500 µg/m²/day) were seen in six of 14 patients with cytopenias resulting from malignancy (either due to marrow infiltration or following therapy).[10] Up to 70-fold increases of the total leucocyte counts were recorded with minimal toxicity. The potential for reduction of chemotherapy-induced myelotoxicity by GM-CSF has been examined in 16 patients with soft tissue sarcomas in a phase I/II study.[11] Patients received 3–7 days of GM-CSF by continuous intravenous infusion during the phase I part of the trial and were then given a combination of Ifosfamide, Adriamycin and DTIC with subsequent re-introduction of GM-CSF. A second cycle of chemotherapy followed, but without growth factor. There were significantly higher nadir total leucocyte and platelet counts after the first cycle (accompanied by GM-CSF) as compared with the second cycle suggesting a role for this agent in protecting against cytotoxic-induced myelosuppression.

An increasingly employed method of treating patients with refractory or advanced malignancies is to give high doses of chemotherapy followed by autologous marrow rescue. Unfortunately there is often considerable morbidity and mortality associated with this procedure and, at the least, the patient is often in hospital for a lengthy (and costly) period of time until the peripheral blood count returns to normal. A potential role for hemopoietic growth factors could, therefore, be their use to accelerate hemopoietic reconstitution following high dose chemotherapy and/or radiotherapy with bone marrow transplantation. Brandt et al.[12] reported a study of 19 patients with refractory breast cancer or melanoma given high dose combination chemotherapy (cyclophosphamide, carmustine and cisplatinum) followed by autologous bone marrow transplantation. GM-CSF was administered as a continuous intravenous infusion for 14 days starting 3 h after bone marrow infusion. Differing doses of GM-CSF were used in sequential patient groups. The results were compared with those of four consecutively treated historical controls who received the same chemotherapy/transplant regimen. There was accelerated recovery of leucocytes and granulocytes in patients given GM-CSF (although this failed to reach levels of significance — probably because of the small numbers of patients involved). Those patients receiving doses of more than 8 µg/kg/day of GM-CSF had higher leucocyte counts 2 weeks after transplantation and more rapid recovery than those at lower doses but the higher doses were associated with more toxicity. Problems with oedema, weight gain and myalgia occurred at 32 µg/kg and in two patients a 'capillary-leak' syndrome (with acute renal failure in one) occurred. Despite these side effects, the patients receiving GM-CSF generally tolerated the chemotherapy with much less subjective and objective toxicity than controls, with fewer episodes of septicemia and less hepatotoxicity and nephrotoxicity. Three other studies have employed high dose chemotherapy and/or total body irradiation followed by autologous bone marrow transplantation for a total of 36 patients with lymphoid malignancies.[13-15] GM-CSF was given after transplant by daily short infusions[13, 14] or continuous infusion[15] and at doses above 60 μg/m^2/day neutrophil and platelet recovery was significantly accelerated (as compared with historical and concurrent controls). There was also a marked reduction of the days on which patients had fevers and of the duration of hospitalization.

Patients with the Acquired Immunodeficiency Syndrome (AIDS) often have a neutropenia in association with reduced neutrophil function and lymphopenia, a combination which is responsible for the high incidence of life-threatening infections. An important study has investigated the use of GM-CSF in 16 such patients who were leucopenic.[16] Varying doses were used intravenously as an initial 1 h injection followed by a continuous infusion over 14 days. Dose-dependent increases of circulating neutrophils, eosinophils and monocytes were induced. Neutrophil function tests have recently been reported and show that GM-CSF increased responses in the functional assays used for the majority of patients.[17] Eight patients also exhibited increased lymphocyte counts. More prolonged treatment will be necessary in future studies to determine the effects of GM-CSF on morbidity and mortality.

Impressive responses to GM-CSF were seen in 8 patients with the myelodysplastic syndrome[18] given doses of 50–500 μg/m^2/day by continuous intravenous infusion for 2 weeks every month. Dose-dependent increases of absolute neutrophil counts were observed in all patients (5–300-fold increments), as well as increases in eosinophils, monocytes and lymphocytes. Interestingly, rises in platelet and reticulocyte counts were also noted and two patients who had previously required red cell and platelet transfusions became transfusion independent. Bone marrow examination after GM-CSF administration showed reduced blast cell counts and decreases in dysplastic changes in all patients. Recent work from this group, using cytogenetic analysis, has demonstrated that mature granulocytes appearing in the peripheral blood after administration of GM-CSF, were derived not only from normal progenitors but also from the myelodysplastic clone.[19] Two other studies[20, 21] have included a total of 18 patients with myelodysplasia and 12 patients with aplastic anemia. GM-CSF was given as short daily intravenous infusions for 7 day periods only (in both studies, the treatment period was lengthened to 14 days in a small number of later patients). Dose-dependent rises of leucocyte counts (predominantly granulocytes and monocytes) were observed in the majority of patients but only one had an increase of platelets. There was a more pronounced effect in patients with myelodysplasia than in those with aplastic anemia. Unfortunately, the bone marrow blast count increased in four patients requiring discontinuation of GM-CSF.

Granulocyte Colony-Stimulating Factor

G-CSF has not been administered to patients with normal bone marrow reserves in a formal Phase I study, but rather has been used in Phase I/II trials in combination with chemotherapy.[22-24] The Phase I components of these trials demonstrate the activity of this agent in producing rapid increases of the leucocyte counts at doses of 1 μg/kg/day and above. No serious toxicity attributed to G-CSF has been seen. The first Phase I/II study of the use of G-CSF in humans included 12 patients with small cell lung cancer.[22] G-CSF was given by continuous intravenous infusion, initially over 5 days with increasing doses (1–30 μg/kg) being administered to sequential patient groups. This was followed by combination chemotherapy using Etoposide, Ifosfamide and Adriamycin. Patients were then assigned sequentially to receive G-CSF on odd or even chemotherapy cycles (commencing 24 h after the end of chemotherapy and continuing for 14 days). Significant rises in total leucocyte counts (predominantly neutrophils) were seen at all dose levels of G-CSF and the count rose within 48 h of commencing the infusion. The eosinophilia seen after GM-CSF administration does not occur. The neutrophils were shown to be normal from tests of their phagocytic activity and mobility.[25] Although the depth of the neutrophil nadir following chemotherapy was not always reduced, the period of absolute neutropenia was significantly shortened (by a median of 80%) in those cycles followed by G-CSF with a return to normal (or above normal) counts within 14 days of starting cytotoxic administration (Fig. 3). Importantly, the incidence of severe infections was markedly lower when G-CSF was given. Bone marrow trephines showed a 20% increase in cellularity during G-CSF administration.

A phase I/II study using G-CSF given as twice daily short intravenous infusions at doses of 1–60 μg/kg/day was reported by Morstyn *et al.*[23] G-CSF was given for 5 days before and 10 days after melphalan (25 mg/m^2) to 15 patients with a variety of advanced malignancies. There was a dose-dependent rise (up to 10-fold increase) of neutrophils during the first phase of G-CSF. The duration of neutropenia following melphalan decreased with increasing doses of G-CSF, lasting only 1 day at a dose of 30 μg/kg. Pharmacokinetics performed during this study revealed a biphasic clearance of the growth factor with a t1/2-alpha of 8 min and a t1/2-beta of 110 min.

A third study has shown a protective effect for G-CSF on chemotherapy-

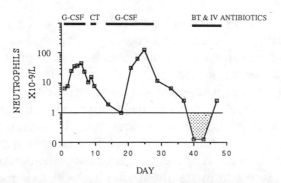

Fig. 3. Profile of neutrophil counts in one patient treated with recombinant G-CSF (as a continuous intravenous infusion) and combination chemotherapy. The time course from the start of therapy (in days) is shown on the lower *x*-axis and the periods during which the patient received the growth factor (G-CSF) at a dose of 10 µg/kg/day, chemotherapy (CT), intravenous antibiotics (IV antibiotics), and blood transfusion (BT) is indicated along the top of the diagram. When chemotherapy was followed by G-CSF the period of absolute neutropenia (count $<1 \times 10^9$/litre) — indicated by the shaded area — was almost completely abolished. In contrast, when the patient received the same chemotherapy without G-CSF, she became severely neutropenic and required admission for intravenous antibiotics and a blood transfusion. (Figure kindly supplied by Dr M. Bronchud from the study in Ref. 22.)

induced myelosuppression.[24] Twenty seven patients receiving doxorubicin, cisplatinum, vinblastine and methotrexate for transitional cell carcinoma of the bladder were given G-CSF (as once daily half-hour intravenous infusions) after the first or second courses of chemotherapy. Administration of growth factor significantly reduced the duration of neutropenia and the number of days on which antibiotics had to be given following cytotoxic therapy. The addition of G-CSF increased the percentage of patients able to receive their subsequent planned course of chemotherapy without delay from 29% to 100% and significantly reduced the incidence of chemotherapy related mucositis.

FUTURE CLINICAL APPLICATIONS

Although only a limited number of clinical studies have been reported to date, it is already apparent that erythropoietin, G-CSF and GM-CSF have important therapeutic indications. Erythroid, leucocyte and, possibly, platelet counts can be increased and these effects have been used to the benefit of patients. The debilitating effects of anemia associated with a variety of diseases can be corrected by erythropoietin. The duration of leucopenia and toxicity of chemotherapy can be decreased by both G-CSF and GM-CSF. The optimal scheduling and route of administration of erythropoietin have not yet been determined but it appears that G-CSF and GM-CSF should be given by continuous intravenous infusion or subcutaneous injection at doses of between 3 and 10 µg/kg/day to obtain the maximal biological effect.

Further studies are now necessary to determine whether the dosages of chemotherapeutic agents can be increased and/or their scheduling can be intensified by safely reducing myelosuppression, the previous major dose-limiting toxicity. Both manoeuvres may, theoretically, be expected to improve response rates and survival for patients with a variety of neoplasms. Another area of potential clinical application of hemopoietic growth factors is to administer them after high-dose chemotherapy and/ or radiotherapy with allogeneic bone marrow transplantation. Such allogeneic transplants are increasingly used for patients with hematological malignancies (particularly leukemias and lymphomas) and the addition of growth factors should reduce the time to full engraftment. Unfortunately other toxicities of chemotherapy, e.g. mucositis, cardiotoxicity, renal impairment, will remain and may become dose limiting. The observation of tumor regression in a phase I study of GM-CSF[7] increases the potential value of this agent in future studies when it is combined with cytotoxic drugs.

An area of considerable interest for the future use of myeloid growth factors is in patients with acute and chronic myelogenous leukemias. There is in-vitro evidence that use of these agents could be contraindicated in such patients — leukemic populations are dependent on the CSFs for proliferation, the quantitative responsiveness being similar to that of normal cells.[26] In addition, two of the CSF genes have been demonstrated to act as proto-oncogenes capable of transforming murine preleukemic cells to fully leukemic cells.[27] On the other hand, GM-CSF and G-CSF may induce differentiation commitment in responding neo-plastic populations. For example, exposure of WEHI-3B leukemia cell lines to G-CSF leads to differentiated post-mitotic granulocytes and monocytes.[28] More importantly, the self-generation to myeloid leukemic stem cells can be suppressed by G-CSF *in vitro* or in animals transplanted with leukemic cells and eventually the leukemic population is completely extinguished by exposure to this growth factor.[29] Similar results are obtained using GM-CSF or G-CSF in human promyelocytic leukemia cell lines.[30] Therefore, although there remains the concern that administration of CSFs to patients with leukemia could accelerate the progress of their disease, there is the intriguing possibility that these agents may have the opposite effect and suppress leukemic populations by favoring differentiation over self-renewal. Anecdotal evidence exists to suggest that it may be reasonable to be optimistic about the effect which myeloid growth factors could have. Buchner *et al.*[31] administered

GM-CSF to six patients with acute leukemia who were at high risk of death and were aplastic following chemotherapy and noted a more rapid recovery of neutrophil count than in controls. There was no evidence for stimulation of the leukemic population in any of the patients. An alternative approach to the use of myeloid growth factors in leukemia could be to combine them with chemotherapy as there is in-vitro evidence that GM-CSF enhances the sensitivity of leukemic clonogenic cells to long-term low dose cytosine arabinoside with sparing of normal clonogenic cells.[32] Such an approach may be safer, at least in initial studies, than giving CSFs alone.

The majority of studies with hemopoietic growth factors have involved patients with malignant diseases. Studies in nonneoplastic conditions (other than renal failure) are still at an early stage but it must be a priority to develop the initial promising work with GM-CSF in patients with AIDS. It will be important to explore the potential for these agents to reduce the morbidity and mortality of patients whose immune function is suppressed as a result of major trauma (for example, to prevent or reduce the severity of infections which frequently occur in burns victims or in patients who have undergone major surgery). G-CSF and GM-CSF could also be used for patients with aplastic anemia or familial neutropenias during episodes of infection — this might be a more appropriate use of these agents than giving them over a prolonged period, at least at the present time.

The imminent availability for clinical testing of Interleukin-3 and M-CSF raises the possibilities of potentiating the effects of those factors currently available by administering them in combination. Interleukin-3 increases the eosinophil, platelet and reticulocyte counts in primates and a synergistic effect is noted if GM-CSF is infused after the IL-3.[33] High dose erythropoietin could be combined with IL-3 and/or GM-CSF to promote increases of leucocytes, platelets and erythroid cells in patients with pancytopenias.

The normal in-vivo response to infection is the elevation of the levels of more than one type of CSF. It would therefore seem logical to assume that optimal results in terms of the prevention of sepsis in humans could be achieved by administering two or more CSFs and a combination of G-CSF and GM-CSF would be one sensible choice. In view of the influx of mature cells to the site of injection local application of growth factors to localized infected areas (for example, peritoneum) may be considered. Although only limited data are at present available on the in-vivo effects of combinations of hemopoietic growth factors,[34, 35] or the combined effects of these plus other bioregulatory molecules (such as IL-1, tumor necrosis factor, transforming growth factor and others), in-vitro data leads to the expectation that combinations of factors may be tailored to specific therapeutic needs. The need to rapidly expand the numbers of mature functional cells (for example, after chemotherapy or radiotherapy, or to combat infection) would indicate the use of GM-CSF or G-CSF, while the need to expand the primitive cell populations (for example to correct aplastic anemia), may require the use of IL-3 or other factors which will preferentially stimulate the stem cells. These choices will be dictated by the windows of action of the growth factors and their specific effects on cell proliferation and differentiation.

CONCLUSIONS

Although many questions remain about the clinical use of hemopoietic growth factors, it is remarkable how much information has been gained in such a short time. There is enormous interest in these agents among physicians from a variety of specialties — all providing new ideas for potential studies. As the supplies increase, many new trials can be opened. It is important to design these studies with due regard to the results of in-vitro and in-vivo preclinical experiments — preferably after discussion with experimental hematologists — or vital information will be missed. There can have been few more exciting developments in Clinical Medicine in the recent past than the availability of the hemopoietic growth factors and, hopefully, it will not be long before the preliminary trials are completed and these agents are accepted for routine use to the benefit of a large number of patients.

REFERENCES

1. Winearls, C. G., Pippard, M. J., Downing, M., Oliver, D. O., Reid, C. & Cotes, P. M., Effect of human erythropoietin derived from recombinant DNA on the anemia of patients maintained by chronic hemodialysis. *Lancet,* **ii** (1986) 1175–8.
2. Esbach, J. W., Egrie, J. C., Downing, M. R., Browne, J. K. & Adamson, J. W., Correction of the anemia of end-stage renal disease with recombinant human erythropoietin. Results of a Phase I and II clinical trial. *N. Engl. J. Med.,* **316** (1987) 73–8.
3. Adamson, J. W., Egrie, J. C., Browne, J. K., Downing, M. R. & Esbach, W., The use of recombinant human erythropoietin (EPO) to correct the anemia of end-stage renal disease: A progress report. *Behring Inst. Mitt.,* **83** (1988) 188–92.
4. McDonald, T. P., Cottrell, M. B., Clift, R. E., Cullen, W. C. & Lin, F. K., High

doses of recombinant erythropoietin stimulate platelet production in mice. *Exp. Hematol.,* **15** (1987) 719–21.

5. Herrmann, F., Schulz, G., Lindemann, A., Meyenburg, W., Oster, W., Krumwieh, D. & Mertelsmann, R., Yeast expressed Granulocyte-Macrophage Colony-Stimulating factor in cancer patients: A phase Ib clinical study. *Behring Inst. Mitt.,* **83** (1988) 107–18.

6. Rifkin, R. M., Hersh, E. M. & Salmon, S. E., A phase I study of therapy with recombinant Granulocyte-Macrophage Colony-Stimulating factor administered by IV bolus or continuous infusion. *Behring Inst. Mitt.,* **83** (1988) 125–33.

7. Steward, W. P., Scarffe, J. H., Austen, R., Crowther, D. & Loynds, P., Phase I study of recombinant DNA Granulocyte-Macrophage Colony-Stimulating factor (rGM-CSF). *Proc. Am. Soc. Clin. Oncol.,* **7** (1988) 614.

8. Morstyn, G., Lieschke, G., Cebon, J., Maher, D., Villeval, J. L., Duhrsen, U., McGrath, K., Boyd, A., O'Connor, M., Nicola, N. A., Green, M., Sheridan, W., Metcalf, D., Fox, R., Rallings, M., Spiegel, R. & Bonnem, E., Phase I study of bacterially-synthesised Granulocyte-Macrophage Colony-Stimulating Factor. Recent developments with hematopoietic growth factors and Intron A. XII Congress Int. Soc. Hematol., Milan, 1988, Abstract 6.

9. Metcalf, D., *The Hemopoietic Colony Stimulating Factors.* Elsevier, Amsterdam, 1984.

10. Vadhan-Raj, S., Buescher, S., LeMaistre, A., Keating, M., Walters, R., Ventura, C., Hittelman, W., Broxmeyer, H. E. & Gutterman, J. U., Stimulation of hematopoiesis in patients with bone marrow failure and in patients with malignancy by recombinant human Granulocyte-Macrophage Colony-Stimulating Factor. *Blood,* **72** (1988) 134–41.

11. Antman, K., Griffin, J. & Elias, A., Effect of rGM-CSF on chemotherapy induced myelosuppression in sarcoma patients. *Proc. Am. Soc. Clin. Oncol.,* **7** (1988) 160.

12. Brandt, S. J., Peters, W. P., Atwater, S. K., Kurtzberg, J., Borowitz, M. J., Jones, R. B., Shpall, E. J., Bast, R. C., Colleen, J. G. & Oette, D. H., Effect of recombinant human granulocyte-macrophage colony-stimulating factor on hematopoietic reconstitution after high-dose chemotherapy and autologous bone marrow transplantation. *N. Engl. J. Med.,* **318** (1988) 869–76.

13. Nemunaitis, J., Singer, J. W., Buckner, C. D., Hill, R., Storb, R., Thomas, E. D. & Appelbaum, F. R., Use of recombinant human granulocyte-macrophage colony-stimulating factor in autologous marrow transplantation for lymphoid malignancies. *Blood,* **72** (1988) 834–6.

14. Blazar, B. R., Widmer, M. B., Kersey, J. H., Ramsay, N. K. C., McGlave, P. B., Urdal, D. L., Gillis, S., Henney, C. & Vallera, D. A., Recombinant human granulocyte-macrophage colony-stimulating factor in human and murine bone marrow transplantation. *Behring Inst. Mitt.,* **83** (1988) 170–80.

15. Link, H., Freund, M., Kirchner, H., Stoll, M., Schmid, H., Bucksy, P., Seidel, J., Schulz, G., Schmidt, R. E., Riehm, H., Poliwoda, H. & Welte, K., Recombinant human granulocyte-macrophage colony-stimulating factor (rhGM-CSF) after bone marrow transplantation. *Behring Inst. Mitt.,* **83** (1988) 313–19.

16. Groopman, J. E., Mitsuyasu, R. T., DeLeo, M. J., Oette, D. H. & Golde, D. W., Effect of human granulocyte-macrophage colony-stimulating factor on

myelopoiesis in the Acquired Immunodeficiency Syndrome. *N. Engl. J. Med.,* **317** (1987) 593–8.

17. Baldwin, G. C., Gasson, J. C., Quan, S. G., Fleischmann, J., Weisbart, R., Oette, D., Mitsuyasu, R. T. & Golde, D. W., Granulocyte-macrophage colony-stimulating factor enhances neutrophil function in acquired immunodeficiency syndrome patients. *Proc. Nat. Acad. Sci., USA,* **85** (1988) 2763–6.

18. Vadhan-Raj, S., Keating, M., LeMaistre, A., Hittelman, W. N., McCredie, K., Trujillo, J. M., Broxmeyer, H. E., Henney, C. & Gutterman, J. U., Effects of human granulocyte-macrophage colony-stimulating factor in patients with myelodysplastic syndromes. *N. Engl. J. Med.,* **317** (1987) 1545–52.

19. Hittelman, W. N. & Vadhan-Raj, S., Cytogenetic evidence for *in vivo* maturation of the abnormal clone after treatment of patients with recombinant human granulocyte-macrophage colony-stimulating factor (GM-CSF). 3rd Conference on Differentiation Therapy, Sardinia, 1988, Abstract 97.

20. Hoelzer, D., Ganser, A., Greher, J., Volkers, B. & Walther, F., GM-CSF in patients with myelodysplastic syndromes — a Phase I/II study. 3rd Conference on Differentiation Therapy, Sardinia, 1988, Abstract 57.

21. Antin, J. H., Smith, B. R, Holmes, W. & Rosenthal, D. S., Phase I/II study of recombinant human granulocyte-macrophage colony-stimulating factor (GM-CSF) in aplastic anemia and myelodysplastic syndrome. *Blood,* **72** (1988) 705–13.

22. Bronchud, M. H., Scarffe, J. H., Thatcher, N., Morgenstern, G., Crowther, D., Souza, L. M., Alton, N. K., Testa, N. G. & Dexter, T. M., Phase I/II study of recombinant human granulocyte colony-stimulating factor in patients receiving intensive chemotherapy for small cell lung cancer. *Brit. J. Cancer,* **56** (1987) 809–13.

23. Morstyn, G., Campbell, L., Souza, L. M., Alton, N. K., Keech, J., Green, M., Sheridan, W., Metcalf, D. & Fox, R., Effect of granulocyte colony-stimulating factor on neutropenia induced by cytotoxic chemotherapy. *Lancet,* **i** (1988) 667–72.

24. Gabrilove, J. L., Jakubowski, A., Scher, H., Sternberg, C., Wong, G., Grous, J., Yagoda, A., Fain, K., Moore, M. A. S., Clarkson, B., Oettgen, H. F., Alton, K., Welte, K. & Souza, L., Effect of granulocyte colony-stimulating factor on neutropenia and associated morbidity due to transitional cell carcinoma of the urothelium. *N. Engl. J. Med.,* **318** (1988) 1414–22.

25. Bronchud, M. H., Potter, M., Morgenstern, G., Blasco, M. J., Scarffe, J. H., Thatcher, N., Crowther, D., Souza, L. M., Alton, N. K., Testa, N. G. & Dexter, T. M., *In vitro* and *in vivo* analysis of the effects of recombinant granulocyte colony-stimulating factor in patients. *Brit. J. Cancer,* **58** (1988) 64–9.

26. Metcalf, D., The molecular biology and functions of the granulocyte-macrophage colony-stimulating factors. *Blood,* **67** (1986) 257–67.

27. Lang, R. A., Metcalf, D., Gough, N. M., Dunn, A. R. & Gonda, T. J., Expression of a hematopoietic growth factor cDNA in a factor-dependent cell line results in autonomous growth and tumorigenicity. *Cell,* **45** (1985) 531–42.

28. Metcalf, D., & Nicola, N. A., Autoinduction of differentiation in WEHI-3B leukemia cells. *Int. J. Cancer,* **30** (1982) 773-7.

29. Metcalf, D., Regulator-induced suppression of myelomonocytic leukemia cells: clonal analysis of early cellular events. *Int. J. Cancer,* **30** (1982) 203-7.

30. Begley, C. G., Metcalf, D. & Nicola, N. A., Purified colony-stimulating factors (G-CSF and GM-CSF) induce differentiation in human HL-60 leukemic cells with suppression of clonogenicity. *Int. J. Cancer,* **39** (1987) 99-106.

31. Buchner, T. H., Hiddemann, W., Zuhlsdorf, M., Koenigsmann, M., Bockmann, A., van de Loo, J. & Schulz, G., Human recombinant granulocyte-macrophage colony-stimulating factor (GM-CSF) treatment of patients with acute leukemias in aplasia and at high risk of early death. *Behring Inst. Mitt.,* **83** (1988) 308-12.

32. deWitte, T., Muus, P., Haanen, C., van der Lely, N., Koekman, E., van der Locht, A., Blankenborg, G. & Wessels, J., GM-CSF enhances sensitivity of leukemic clonogenic cells to long-term low dose cystosine arabinoside with sparing of the normal clonogenic cells. *Behring Inst. Mitt.,* **83** (1988) 301-7.

33. Donahue, R. E., Seehra, J., Norton, C., Turner, K., Metzger, M., Rock, B., Carbone, F., Sieghal, R., Young, Y. C. & Clark, S., Stimulation of hematopoiesis in primates with human interleukin-3 and granulocyte-macrophage colony-stimulating factor. *Blood,* **70** (1987) 133a (Abstract).

34. Broxmeyer, H. E., Williams, D. E, Hangoc, G., Cooper, S., Gillis, R. K., Shadduck, R. K. & Bicknell, D. C., Synergistic myelopoietic actions *in vivo* after administration to mice of combinations of purified natural murine colony-stimulating factor-1, recombinant murine interleukin-3, and recombinant murine gramulocyte-macrophage colony-stimulating factor. *Proc. Nat. Acad. Sci., USA,* **84** (1987) 3871-5.

35. Broxmeyer, H. E., Williams, D. E., Cooper, S., Shadduck, R. K., Gillis, S., Waheed, A., Urdal, D. L. & Bicknell, D. C., Comparative effects *in vivo* of recombinant interleukin-3, natural murine colony-stimulating factor-1, and recombinant murine granulocyte-macrophage colony-stimulating factor on myelopoiesis in mice. *J. Clin. Invest.,* **79** (1987) 721-30.

Chapter 16

SELECTIVE DELIVERY AND TARGETING OF THERAPEUTIC PROTEINS

E. TOMLINSON

Advanced Drug Delivery Research, Ciba-Geigy Pharmaceuticals, Horsham, West Sussex, UK

INTRODUCTION

The advances made during the past decade in recombinant DNA science has led to the ability to identify, clone, express and produce on a large scale many proteins. As more becomes known about the pharmacology of proteins and polypeptides, it is becoming increasingly apparent that a variety of new peptide and protein drugs can be expected to be used clinically in the coming decades. Polypeptides and proteins proposed for therapy usually have regulatory or homeostatic functions. They include both endogenous polypeptides and proteins and their (heterologous) derivatives. This latter class of molecules may be produced after site-directed mutagenesis or gene fusion, or by proteolysis or protein aggregation, and/or conjugation with (other) biologically active effector functions.

The most widely used method of administration of conventional low molecular weight drugs is via the oral route. It is however unlikely that this route will be a feasible one for the vast majority of polypeptides and proteins, due to their degradation by proteases in the gastrointestinal tract, and their inability to traverse epithelial membranes. Thus, although most of the putative therapeutic proteins are intended for chronic use, polypeptides and proteins will generally be administered parenterally. Once within the body, low molecular weight drugs usually move to site(s) of action via passive means — their distribution being controlled by their physicochemistry and lability. However, the macro-molecular nature of proteins gives a restriction to such passage between tissue and organ compartments. In addition, the regulatory and/or homeostasic functions of proteins often require them to be available to their site(s) of action(s) on very precise occasions, and generally in reponse to time-related changes in biological status. The use of various protein engineering and recombinant DNA techniques have enabled progression to where proteins can be tailored so that they have not only the ability to interact with a unique (extra)cellular component in order to produce a pharmacological effect, but their structure can be altered so as to achieve a *selective biological disposition*. This present chapter examines some of the features of protein (physico)chemistry and pharmacology which determine the format of delivery required to achieve appropriate spatial and temporal arrival of polypeptides/proteins at their site(s) of action.

GENERAL CONSIDERATIONS

With most therapeutic proteins there is a highly critical relation between applied dose and effect. Many even exhibit peculiar non-linear dose–effect relationships (for example with parathyroid hormone, substance P and δ-sleep inducing peptide). The administration pattern of a therapeutic polypeptide or protein is often a strong determinant of its resultant pharmacokinetic and pharmacodynamic behaviour. For example, Robinson and his colleagues have examined the growth responses to intravenous growth hormone administered to hypophysectomised animals as various pulsitile infusions.[1] They were able to show that pulsitile growth hormone produced a greater effect on animal growth with time than did (unphysiological) continuous infusions. It is apparent that many putative therapeutic polypeptides and proteins are administered in a manner which either does not mimic physiologic delivery patterns, and/or are inappropriate for the biological process which they are intended to modify.

A large number of proteins intended for therapeutic use are glycoproteins, whose biological disposition is primarily due to three properties: chemical and metabolic stability, size and shape, and surface features.

Biological Stability

Most polypeptides and proteins have a short biological half-life, which is often due to a poor chemical stability and/or rapid liver metabolism and kidney excretion. Paracrine-like and autocrine-like mediators are often unstable. This is largely due to their degradation by peptidases and proteinases in the vascular endothelium, liver, and kidneys, etc. The terminal amino acids of a protein chain can serve to control *intra-cellular* metabolic stability — and hence intracellular residence time.[2] Protein-engineering methods may be used to replace labile amino acids — such as the oxidation-resistant amino acids alanine, serine or threonine, or to produce proteins having differing foldings — potentially leading to proteins which are protected from inactivation.

Size and Surface Character

Therapeutic (glyco)proteins can be both flexible and/or globular, and be up to 300 kD in size in the non-aggregated state. This affects their ability to diffuse across endo- and epithelial membranes. Analysis of the (patho)physiological and anatomic opportunities and constraints for the movement of macromolecules within the body, shows that membrane size selectivities are highly dependent upon the region of tissue and/or any pathology.[3] For example, when placed into the vasculature, macromolecules above 30 nm diameter are restrained within that compartment, and hence are able to interact with the surfaces of blood and endothelial cells, and also with exposed parenchymal cells of organs having discontinuous endothelia (e.g. liver and spleen).

Polymeric State

The polymeric state of a protein may affect its pharmacodisposition. This is exemplified by recent work on monomeric insulin produced by protein engineering methods.[4] This form of insulin is found to be absorbed into the body two to three times faster than its polymeric form after subcutaneous administration.

PHARMACOLOGY OF THERAPEUTIC PROTEINS

Most conventional low molecular weight drugs used chronically need to be available at their site of action at a constant amount over a constant period. However, for therapeutic polypeptide and proteins, it is often the case that arrival at their site of action needs to be asymmetric with relation to time.

Site of Action

Current approaches to the development of protein drugs have given little rational consideration to their arrival at their site(s) of action(s). That is, should the protein be acting systemically (i.e. endocrine-like), or it is a mimic of an endogenous molecule that is normally produced to act locally (i.e. autocrine/paracrine-like). For the parenteral administration of endogenous proteins (e.g. hormones such as insulin) that would normally circulate in the blood, little problem arises. Such proteins are produced naturally to act over long distances from their site of production; they are also stable in blood, and, if relevant, their size and surface character enable their (specific) extravasation. However, the autocrine/paracrine-like mediators are produced and released in order to act locally, and/or have very short chemical half-lives. Such properties ensure that they do not give rise to untoward effects on non-target neighbouring cells. This latter class of mediators are often produced at sites of inflammation, tumours and injuries (e.g. transforming growth factors alpha and beta, angiogenin, fibroblast growth factors, and epidermal growth factors, etc.). Examine, for example, the properties and suggested uses of growth factors. This diverse group of polypeptide hormones regulates cell function and metabolism, and is active in tissue repair. Hence, they are suggested for the treatment of burns and trauma, etc. Growth factors may be endocrine-like or paracrine/autocrine-like mediators. For the latter class, their specificity of action is due to their ability to interact with specific cell receptors on the surface of near-neighbour cells, and/or their chemical instability in blood. Most of these growth factors often exert their cellular action in conjunction with additional cofactors. Their mechanism of action includes binding to a cell surface membrane receptor followed by enzyme activation leading to receptor phosphorylation. This triggers cellular mitosis (DNA synthesis and cellular division), often after some hours. In addition, and importantly, oncogenes produce protein products which are functionally homologous to many growth factors. Since the cell surface interactions of growth factors can have large cross-reactivities with other (related) surface receptors, and because of the clear interrelation with oncogene products, the use of such molecules appears to be contraindicated unless means can be found for controlling their delivery to their sites of action in a manner which mimics their (exclusive) endogenous delivery. Also, some growth factors may be classed as endocrine-like. For example, the insulin-like growth factor IGF1, is produced in the liver, and requires an endogenous carrier protein to transport it to its site of action. For growth factors, the numerous interplays between cell specificity, chemical stability, requirement for carrier proteins, local and/or systemic action and cellular specificity, are complex, and provide obvious challenges to the achievement of their correct delivery.

Hence, for autocrine/paracrine-like mediators it can often be the case that if administered parenterally, because of their size and/or chemical/

metabolic stability, they will be unable to reach an extravascular intra- or extracellular site of action. In addition, since these types of molecules can produce different effects on the same cells when these are at different stages of their life cycle, their delivery should be formulated to follow an appropriate biological rhythm.

ADMINISTRATION

The large majority of polypeptide and protein drugs will need to be administered parenterally and/or interstitially, until innovative approaches can be developed that can enable their successful transepithelial transport. Some opportunities do exist for their administration via routes that are more convenient to patient and physician alike.[5] (It has been considered unlikely that this would be possible with administration to *non-mucosal* barriers, though interestingly, recent reports[6] claim that insulin can be administered transdermally using electrical current as an inducer.)

Parenteral/Interstitial

Apart from parenteral administration of endocrine-like therapeutic proteins both as solutions or in long-acting dosage forms, opportunities exist for the interstitial administration of therapeutic proteins. Numerous groups are developing controlled-release formulations which would enable varied forms of protein release from implants. Biodegradable polymers have been studied to obtain a constant and prolonged input of therapeutic protein (e.g. Ref. 7).

Oral

The physical and chemical properties of therapeutic proteins make them unlikely candidates for oral administration. That is, they are large, are often chemically unstable — and hence degrade in the environment of the lumen of the gastrointestinal tract — and they have a poor ability to pass biological membranes via either a passive transcellular or a paracellular route. Numerous attempts are being made to overcome each of these difficulties, with much current attention being given to those which are rate-controlling (e.g. Ref. 8). The epithelial membrane of the gastrointestinal tract comprises an anatomically continuous barrier of cells, which permits the passage of low molecular weight material by simple diffusion and various (nutrient) carrier processes. Macromolecules may be absorbed from the lumen by cellular vesicular processes via either fluid-phase pinocytosis or specialised (receptor-mediated) endocytic processes. Also, specialised cells (M cells) exist within the Peyer's patches of the gastrointestinal tract epithelium that can take up soluble macromolecules and transport these through into the lymph. Although the capacity of this system to transport macromolecules appears to be extremely low, since mucosal immune responses are initiated within this lymphoidal tissue, it may be that the M cell could be used clinically to bring an appropriate antigen, e.g. an immunomodulatory protein, to the attention of the mucosal immune system.

The epithelial absorption of macromolecules may be enhanced using various adjuvants. These include water–oil–water multiple emulsions, ionic and non-ionic surfactants, mixed bile salt micelles, surfactant–lipid mixed micelles, etc., all of which may cause a transient reduction in the resistivity of the surface of the apical membrane of the epithelial cells. For example, sodium glycocholates or polyethylene-9-dodecyl ether enhances the transmucosal permeation of insulin.[9] Similarly, a marked increase in plasma levels of human epidermal growth factor (EGF) is observed when EGF is administered rectally as a microenema containing hydrophilic polymers such as sodium caprolate and sodium lauroylphenylalanine.[10] Other enhancers include salicylates (for growth hormone).[11] Studies on the enteric absorption of α- and β-human interferons in the rat large intestine have shown that addition of apricot kernel oil and polyethylene glycol 6 to a citrate-buffered solution favours the blood bioavailability of α-, but not of β-interferon. This could be due to an increase in both the stability of α-interferon, and/or in its penetration.[12] Recent work has shown that small amounts of interferon-α_2 are absorbed from the oropharynx in rats using sodium ursodeoxycholate as absorption promotor.[13]

Nasal

Recent attention has been given to the *nasal* administration of macromolecular polypeptide and protein drugs. (For example, insulin and calcitonin have been found to be pharmacologically active after being applied to the nasal mucosa.) However, for most large polypeptides and proteins, molecular size restricts their absorption across the nasal mucosa. Also, polypeptides appear to undergo rapid hydrolysis in the nasal cavity. These factors are leading to numerous approaches to improving on the nasal absorption of polypeptides and proteins. That is, the use of competitive peptides or highly concentrated solutions of peptides and/or inhibitors of peptidase enzymes (e.g. puromycin); the use of permeation enhancers such as fusidates, and cholic acids;

increasing blood flow; and prolongation and controlled release of proteins at the nasal mucosa.[14] Present evidence suggests that nasal administration for polypeptides results in bioavailabilities of between 1 and 20% of applied dose, depending on their molecular weight and physicochemical properties. However, nasal administration of poly-peptides and proteins may lead to tolerance, toxic events and unique immune responses (e.g. Ref. 15).

DELETION MUTANTS

Site-directed mutagenesis can be used to create *de novo* heterologous proteins which may or may not broadly resemble endogenous material. These approaches are being used not only to improve on the stability and intrinsic specificity of such endogenous proteins, but also (and increasingly), to achieve a selective and often prolonged delivery of the polypeptide/protein to an active site. For example, recent work has demonstrated the altered pharmacokinetic and thrombolytic properties of deletion mutations of human tissue-type plasminogen activator (t-PA) in rabbits.[16] Wild-type t-PA is characterised by a rapid clearance by the liver, with an alpha distribution phase half-life of a few minutes. Using a series of deletion mutants (which included removal of fibronectin and epidermal growth factor domains and glycosylation sites) this group have demonstrated that regions within t-PA responsible for its liver clearance, its fibrin affinity, and its fibrin specificity are not localised in the same structures. They argue that it appears possible to alter specific functions of tPA related to poor pharmacokinetics without decreasing its efficacy. Haber *et al.*[17] have recently reviewed how the tools of molecular biology and protein engineering may be used to develop 'safer and more effective' plasminogen activators. They describe both domain-deletion and site-directed mutagenesis techniques for the creation of new plasminogen activators, as well as chimaeric (or hybrid — see below) molecules. Haber *et al.* review the use of domain deletions to produce a shortened form of single-chain urokinase-like plasminogen activator (scuPA), which does not have the NH_2-terminal domain kringle structure of scuPA.

HYBRID PROTEINS

Knowledge of the structure, position and function of many of the operational receptors in the body in controlling the extracellular and intracellular disposition of proteins, is leading to the design of heterologous *hybrid* proteins which have the combined or re-ordered features of one or more proteins, in order to have both effector functions as well as protection and recognition properties. Intracellular recognition signal structures have been used for designing synthetic peptides able to mimic (secretory) events (e.g. Ref. 18). Oligonucleotide-directed mutagenesis, etc., to produce hybrid proteins is becoming common in elucidating cell function (e.g. the intracellular dispersion of secretory proteins[19]). Site-specific hybrid proteins may be produced by either synthetically linking protein fragments (e.g. Ref. 20), or using ligated gene fragments followed by expression. Table 1 gives examples of hybrid proteins which have been either produced and/or suggested for therapeutic use.

Table 1
Hybrid Protein Delivery Systems

Fragments of recognition portion	*Fragments of effector portion*
Gene fusion products	
Interleukin-2	Diphtheria toxin
Growth factor	Toxin (e.g. ricin A chain)
Cell-specific polypeptide	
(α/β MSH; substance P)	Restructured diphtheria toxin
Antitumour Fab immunoglobulin	Fragment A of diphtheria toxin
CD_4	*Pseudomonas* exotoxin
	γ-interferon and β-tumour necrosis factor (mechanism of recognition/effector function currently unknown[23]
Chemical linkage of fragments/proteins	
Human placental lactogen hormone	Diphtheria toxin A chain
β chain of *h* chorionic gonadotropin	Ricin A chain
Insulin	Diphtheria A chain
Epidermal growth factor	Ricin A chain
Antibody fragments	(Deglycosylated) Ricin A chain
IgG(2a) fragments	Gelonin toxin
	Diphtheria toxin
	Pseudomonas toxin
Anti-collagen antibody	Toxin
HIV-specific Ab	Ricin A chain
Antifibrin Ab	Tissue Plasminogen Activator
Anti-T cell antibody	Ricin A chain
Antibody fragments	Gelonin toxin
Anti-endothelia IgG	Glucose oxidase
Anti-epithelia Ig/fragments	Ricin and other toxins
Anti-sIgM-IgM	Saporin-6
Anti-EGF-receptor Ab	Ricin A chain[28]
Anti-transferrin receptor Ab	Ricin A chain

Ligated Gene Fusion Hybrids

Gene-fusion techniques may be used to produce distinct therapeutic proteins combining the varied properties of parent proteins. This is exemplified by recent strategies which have been proposed for the targeting of bacterial and/or plant toxins to specific cells using hybrids created by ligating toxin and growth factor genes.[21] Their approach relies on the deletion of the toxin gene sequence encoding the cell-binding site, which allows the hybrid-fusion protein to display the cell specificity of the growth factor. Chaudhray *et al.* have also produced CD_4-*Pseudomonas* exotoxin hybrid proteins which show selective toxicity towards cells expressing the HIV envelope glycoprotein which binds to the CD_4 antigen.[22] Recently, a hybrid protein between interferon-γ and tumour necrosis factor-β has been shown to have a greatly increased anti-proliferative activity *in vitro* compared to either interferon-γ or TNF-β alone, whilst still retaining their antiviral activity and cytotoxic effects.[23] As Feng and colleagues point out, the *intramolecular* synergy within the hybrid protein may be via an entirely different process than when the two molecules are applied sequentially.

Again, Haber *et al.*[17] have demonstrated the use of hybrids as novel molecules that combine delivery and effector functions for use in thrombolytic therapy. For example, high-affinity fibrin selectivity and resistance to plasminogen activator inhibitor I can be introduced into a molecule by construction of a hybrid composed of the A chain (fibrin-binding domain) of tissue plasminogen activator, and the low molecular weight form of scuPA (i.e. the catalytic site of urokinase in a form that is not susceptible to plasminogen activator inhibitor). Such enterprise however is not without its difficulties. As Haber *et al.* point out, unfortunately the fibrin-binding activity of the recombinant hybrid is less than that of native tPA, with its fibrin selectivity being found to be less than that of single- or two-chain tPA.

Hybrid protein delivery systems may involve not just adduction of a protein (fragment) to a recognition moiety, but also to a re-ordering of the structure of the effector portions of therapeutic proteins in order to enhance their pharmacological action. For example, Murphy[24] has described a hybrid protein, consisting of (i) the enzymatically active fragment A of diphtheria toxin, (ii) a fragment including the cleavage domain 1_1 adjacent to fragment A, (iii) a fragment which includes (at least) a portion of the hydrophobic domain of fragment B (though not including the generalised eukaryotic binding site of fragment B), and (iv) a fragment which includes a portion of a cell-specific polypeptide able to bind the conjugate delivery system to a specific cell feature joined in sequence by peptide bonds (Table 1).

Included in this category of novel proteins having both recognition and effector functions, are the recently described immunoadhesins.[25] Specifically, these are antibody-like molecules, containing the gp120-binding domain of CD_4, the receptor for human immunodeficiency virus. They have been shown to block HIV-1 infection of T cells and monocytes. Interestingly, such novel hybrids delivery systems have a long plasma half-life. The fusion of the gp120-binding domain of CD_4 to the Fc domain of an immunoglobulin has considerable merit when one appreciates that the formed heterologous protein retains some of the important properties of both of its parent molecules; namely, they bind gp120 and block infection of T cells by the virus. They are also comparable to antibodies in their long plasma half-life, and their ability to bind Fc receptors and protein A. The claims of Capon et al.[25] include that the attainment of a high steady-state level of immunoadhesin makes it likely that effective concentrations of the hybrid will be attained in lymph and lymphatic organs (where HIV may be most active).

Synthetically Linked Hybrid Conjugates

Large changes in the biological disposition of proteins have been reported upon their chemical linkage to other protein (fragments). For example, the toxin gelonin — which has a circulation half-life in mice estimated at 3·5 min when conjugated to immunoglobulin (fragments) has a terminal phase blood half-life in the order of days, with only a slight variation in this time as the conjugated immunoglobulin (fragment) is changed.[26] The immunotoxin field provides many relationships between protein structure and deposition. Greenfield et al.[27] have produced point mutations in the B polypeptide chain of diphtheria toxin that block non-specific binding to non-target cells. Upon linking this entity to an anti-T-cell monoclonal antibody, they demonstrated that because of a change in the non-target tissue distribution of the toxin, it becomes orders of magnitude less toxic than the native toxin to non-target cells (in vitro). The availability of monoclonal antibodies has led many workers to consider these for the targeting of toxic materials to human tumours. Table 1 points to some of the proposed uses of these materials.

A recent and typical example is that of Taetle et al.,[28] who prepared an immunoconjugate from an anti-epidermal growth factor (anti-EGF) receptor antibody and recombinant ricin-A chain. Their data show that this site-specific conjugate gives a dose-dependent killing of cells that express EGF receptors. However, and again typically for these types of anti-proliferative hybrids, the kinetics of cell killing with these conjugates was protracted, suggesting that prolonged exposure may be required for in-vivo anti-tumour effects.

Bispecific antibodies also come into this class of molecules. For example, a bispecific antibody that recognises the epitopes on both fibrin and a plasminogen activator, should be capable of increasing the effective concentration of the plasminogen activator in the proximity of a fibrin deposit.[17]

PROTEIN (RE)GLYCOSYLATION

As described above, the size, surface character and chemical reactivity of proteins largely control their distribution in the body. Numerous groups (e.g. Ref. 29) have demonstrated that many endogenous glycoproteins (i.e. serum glycoproteins, lysosomal enzymes and perhaps also sulphated pituitary glycoproteins such as chorionic gonadotropin), interact through their specific carbohydrate residues complexing with (oligosaccharide-specific) recognition systems on the plasma surfaces of target cells. Glycosylation patterns are thus signals used by the body to regulate the dispersion of its own glycoproteins, as implicated for both enzyme and hormone disposition as well as immune surveillance, coagulation, etc. Three biological properties of glycoproteins may be adjusted upon altering their surface distribution of carbohydrates (a) circulating blood half-life (and potentially duration of action), (b) immunogenicity, and (c) ability to access (cellular) sites of action. What is also intriguing is the recent proposal that carbohydrate structures in transforming growth factor-β1, are also of importance in helping to control the *latency* of this molecule *in vivo*.[30]

Thus, changing the oligosaccharide content of proteins has potential in controlling the fate, and perhaps dynamics of action, of therapeutic glycoproteins. In contrast to possible changes in the amino acid composition of a protein, the numerous variations possible in linking simple sugars together affords *glyco*proteins an almost limitless variability and diversity in structure. Additional modifications, such as removal or addition of peripheral sugars and/or other functional groups such as acetyl, methyl, sulphate and phosphate, are also possible. Oligosaccharides may be N-glycosidically linked (to the peptide at Asn), or O-glycosidically linked (attached to Ser and Thr). The oligosaccharides of the plasma glycoproteins are linked to protein primarily through L-asparagine-N-acetyl-D-glucosamine.

For example, the intracellular translocation of lysosomal enzymes to lysosomes is due to the phospho-D-mannopyranosyl moiety of lysosomal enzymes, such that phosphorylation of the D-mannose residues is essential for their uptake and intracellular transport to lysosomes.[31] Numerous other structure/binding relationships are known. Asialoglycoproteins having oligosaccharides biantennary and triantennary are preferentially adsorbed by leucocyte lectins, whereas those having oligosaccharide triantennary and tetraantennary are more rapidly bound to the plasma membrane of liver hepatocytes. Other studies also indicate that it is the spatial co-relationship between residues, and not their number, that is important in cell recognition. Galactose-specific recognition systems have been identified on hepatocytes (for D-galactose and N-acetyl-D-galactosamine), on lymphocytes, and more recently on macrophages.[32] Similarly, mannose-containing oligosaccharides are taken up by both the endothelial cells and the Kupffer cells of the liver; which leads to the suggestion that reductive *mannosamination* may provide a means of directing therapeutic proteins to such cells.[33] Carbohydrates often play no direct role in the biological activity of the glycoprotein; instead, as structural elements, they may influence the *stability* and *conformation* of glycoproteins. Also, they frequently impart to glycoproteins good aqueous solubility because of their hydrophilic character.

The function of carbohydrate modifications of the natural haematopoietins (including granulocyte-macrophage colony-stimulating factor (GM-CSF), erythropoietin, and CSF-1) have been studied recently.[34, 35] The polypeptide backbone of natural human GM-CSF includes two consensus asparagine-linked glycosylation sites. Extensive modifications produced by the addition of an Asn-linked carbohydrate has been found by Donahue and colleagues to result in rather heterogeneous glycosylation patterns, dependent upon expression cell and protein under study. This group have demonstrated that the effective half-life of a GM-CSF in the bloodstream of a rat is increased significantly by the addition of N-linked carbohydrate. In addition, their recent work has described how novel glycosylation site variants characterised structurally by a reduced presence of carbohydrate moieties relative to fully glycosylated natural (or recombinant GM-CSF), are characterised biologically by an improved specific activity (up to 10-fold higher) relative to fully glycosylated natural or recombinant GM-CSF.[35] Similar claims are given for other granulocyte-macrophage colony-stimulating factors (e.g. Ref. 36).

If galactosyl residues are to be used clinically to target proteins to hepatocytes it is important to appreciate that the in-vivo uptake of galactosylated *neo*glycoproteins has been shown to be non-linear with respect to dose applied.[37]

Expression Cell Processing

Protein glycosylation does not usually occur in prokaryotic cells such as *E. coli*. Some glycosylations do occur in eukaryotes, e.g. yeast and mammalian cells, but the resultant glycosylation patterns can often be different, depending on the cells used. Also, although recombinant bacteria are able to produce large amounts of protein, these are often changed by bacterial proteases; within expression cells it is possible that expressed proteins of low solubility denature and form aggregates. Mammalian cells are particularly suitable both for expressing proteins with complex modifications such as γ-carboxylation of glutamyl residues, and for obtaining homologous (human) glycosylation patterns. Thus, it is apparent that biotechnology processes used to make the protein can have a marked effect on the biological disposition and efficacy of the proteins when used therapeutically.

Protein Re-modelling

Expressed glycoproteins may be engineered in order to affect their selectivity for target cells. Modifications of proteins are usually carried out via enzymatic synthesis. Various chemical approaches to re-modelling proteins post-expression are being developed (e.g. Ref. 38). Recently, an important new approach has been described which employs enzymes to elongate and terminate peripheral glycan chains of glycoproteins.[39, 40] Mammalian glycoproteins expressed in yeasts are likely to be substituted by mannans, and this group has been able to incorporate sialic acid into endo-β-*N*-acetylglucosaminidase H-treated oligomannose glycoproteins.[39] The technology is of potential use in reducing any mannose-receptor uptake of these glycoproteins by cells of the mononuclear phagocyte system, etc. The approach successfully incorporates sialic acid into glycoproteins of the oligomannose type, and hence appears to be promising for the in-vitro re-modelling of glycan chains in heterologous glycoproteins such as tissue plasminogen activator, and α_1-antitrypsin. Post-expression protein re-modelling has been used to resurface some of the enzymes indicated in lysosomal storage diseases. Other proteins reported amenable for this kind of re-modelling include factor VIII, EPO, colony-stimulating factors, β- and γ-interferons, interleukins I-III, and even antibodies.

Deglycosylation

Conversely, protein *de*glycosylation should also affect dispersion (and stability and solubility) of a therapeutic glycoprotein. Indeed, deglycosylation has been shown to virtually eliminate the uptake of the toxin ricin A chain by the liver.[41] Ricin A chain is an oligomannosidic glycoprotein which has a very short half-life due to rapid removal by the mononuclear phagocyte system, and deglycosylation of the A chain may reduce removal by Kupffer cells. Although chemical deglycosylation only marginally delays the clearance of deglycosylated ricin A from the blood compartment (in mice), this is probably due to the production of a smaller macromolecule able to be effectively filtered and excreted through the kidneys. Blakey and Thorpe assume that linking the deglycosylated ricin A chain to an immunoglobulin would diminish this latter effect. Ricin A chain lacking carbohydrate side chains has also been produced using recombinant DNA technology.[42]

HYDROPHILIC PROTECTANTS

Chemical adduction of therapeutic proteins to (hydrophilic) polymers, may be used to alter their biological dispersion. Such conjugation serves to either increase the apparent size of proteins, and/or reduce their (untoward) interactions with blood and tissue components. Immuno-surveillance is mediated through physicochemical interaction between a (therapeutic) protein and components of the immune system. Frequently, *opsonisation* by fibrinogen, fibronectin and other blood components, is a prelude to recognition and thence removal by cells of the formed complex; antigen–antibody interaction and Fc-mediated removal also occurs. Opsonised materials are taken into cells by engulfment after adherence to, and vesiculation of, phagocytosing cell membranes. Both opsonisation and adherence can be diminished if the attractive forces between the interacting therapeutic protein, blood macromolecule and, for example, a cell-surface macromolecule are diminished. Napper and Netschey[43] have argued that for colloids, a high potential energy barrier can be formed by creating a *sterically stabilised* surface upon introducing a hydrated (i.e. hydrophilic) polymer at the surface of the colloid. Hence surface modifications to proteins can be made to improve their *tolerance* within the vasculature — due largely to the formation of a surface which makes it energetically unfavourable for other macromolecules to approach. Table 2 gives examples of hydrophilic (bio)polymeric *protectants* that have been described for conjugation to therapeutic proteins. Synthetic and biological materials have been used or suggested,

Table 2
Protectants for Therapeutic Proteins[a]

Protein	Proposed use of conjugate
Polyethylene glycols	
Islet-activating protein	Insulinogenic activity
Superoxide dismutase	Kidney transplantation, burns, re-perfusion damage
L-Asparaginase	Malignant haematological disorders
Adenosine deaminase	AD deficiency
Urokinase	Anti-coagulant, fibrinolytic
Proteins	Radioprotection
Interleukin-2	Cancer
Uricase	Altered antigenicity
t-PA	Thrombolysis
Bilirubin oxidase	Bilirubinaemia
Catalase	
Interferons	
Insulin	
Growth hormones	
Immunoglobulins	
Trypsin inhibitors	
Proteases and peptidases	
Octylphenoxy polyethoxy ethanols	
Interleukin-2	
Polyoxyethylene sorbitans	
Interleukin-2	
Dextran	
Urokinase	Fibrinolytic
Purine nucleosides	Inhibitors of adenosine deaminase
α-1,4-Glucosidase	
*neo*glycoproteins	Enzyme replacement therapy
Carboxypeptidase G_2	Enzyme replacement therapy
β-Galactosidase	Enzyme replacement therapy
L-Asparaginase	Leukaemia (lower antigen reactivity and increased circulatory persistence)
Superoxide dismutase	
Albumin	
L-Asparaginase	Leukaemia
Poly-D-alanyl peptides	
L-Asparaginase	Leukaemia
N-(2-Hydroxypropyl)methacrylamide (HPMA)	
Antibodies	Cancer seeking agents

[a]Described in the (patent) literature.

and include polyethylene glycols, poloxamers, poloxamines, albumin, immunoglobulin G, carboxymethylcellulose, natural xanthans and sorbitans. Conjugations of proteins with hydrophilic polymers have often been reported as being very successful in altering their *potency* as well as for reducing their immunogenicity and increasing their duration of action. Abuchowski and Davis[44] used this approach for stabilising therapeutic proteins by forming protein conjugates with hydrophilic polyethylene glycol chains. Others have increasingly used this concept, and modifications of it, for extending the blood half-lives of a number of peptidergic mediators such as interleukin-2 (IL-2),[45] and enzymes, including catalase, asparaginase and urokinase, whilst still maintaining their reactive functionalities (Table 2). (It is useful to recall here that to prolong blood levels of polypeptides/proteins is not necessarily desirable, as may be the case with IL-2, since a non-physiologic pattern of exposure to non-target cells may occur.) Significant changes to biological dispersion can be made, as demonstrated by the recent studies of Ho *et al.*,[46] which showed that PEG treated asparaginase in rabbits exhibits an increase in its circulating half-life over the native enzyme, of from about 20 h to between about 125 h and 160 h; with a greatly reduced clearance from the blood of from about 100 ml/kg/day to about 3 ml/kg/day, and with an almost 40-fold increase in the plasma area under the curve. It has been reported that an asparaginase conjugate is well tolerated in patients who had already been treated with native asparaginase and had neutralising anti-asparaginase antibodies. The approach is demonstrated as being useful for prolonging blood residence times and improving on immunogenicity. However effects on alteration in potency are occasionally somewhat less encouraging.

CONCLUSION

(Physico)chemical structure, and routes and pattern of administration, can affect the pharmacodynamics, pharmacokinetics and toxicity of therapeutic proteins. The first of the homologous proteins able to be successfully developed as therapeutic agents, will be those which are mainly to act intravascularly at extra- and/or intracellular sites, and where the protein is able to access tissue parenchyma through discontinuous endothelia. In order to use clinically future generations of protein drugs (e.g. autocrine/paracrine-like mediators, etc.), it is apparent that their (physico)chemical structure, and route and pattern of administration, will need to be such as to ensure that their delivery to their sites of action is *physiologically relevant*. Structural changes may be brought about by, *inter alia*, deletion mutation, hybridisation, (re)glycosylation and/or chemical adduction. These will result in changes in protein stability, site recognition and binding, and disposition. Additionally, it will be necessary to know whether symmetrical and/or asymmetrical administration patterns are required. Such patterns will be achieved by giving such substances via the correct route, and at an appropriate amount, rate, frequency, and duration, as well as with a proper staging of application with (other) (proteinaceous) mediators.

REFERENCES

1. Clark, R. G., Jansson, J.-O., Isaksson, O. & Robinson, I. C. A. F., Intravenous growth hormone: growth responses to patterned infusions in hypophysectomized rats. *J. Endocr.,* **104** (1985) 53–61.
2. Bachmair, A., Finley, D. & Varshavsky, A., In vivo half-life of a protein is a function of its amino-terminal residue. *Science,* **234** (1986) 179–86.
3. Tomlinson, E., Theory and practice of site-specific drug delivery. *Advanced Drug Delivery Reviews,* **1** (1987) 87–198.
4. Brange, J., Ribel, U., Hansen, J. F., Dodson, G., Hansen, M. T., Havelund, S., Melberg, S. G., Norris, F., Norris, K., Snel, L., Sørensen, A. R. & Voigt, H. O., Monomeric insulins obtained by protein engineering and their medical implications. *Nature,* **333** (1988) 679–82.
5. Epstein, D. A. & Longenecker, J. P., Alternative systems for peptides and proteins as drugs. *CRC Crit. Rev. Therapeut. Drug Carrier Systs,* **5** (1988) 99–139.
6. Meyer, B. R., Katzeff, H., Eschbach, J. C., Trimmer, J., Zacharias, S. & Rosen, S., Successful transdermal delivery of human insulin to rabbits with alloxan-induced diabetes mellitus. *Clin. Res.,* **36** (1988) 367A.
7. de Nijs, H., Bouwman, T. R. M. & Eenink, M. J. D., Controlled peptide delivery using biodegradable microcapsule formulations. *Pharm. Weekblad Sci. Ed.,* **10** (1988) 49.
8. Matuszewska, B., Liversidge, G. G., Ryan, F., Dent, J. & Smith, P. L., In vitro study of intestinal absorption and metabolism of 8-L-arginine vasopressin and its analogues. *Int. J. Pharmaceutics,* **46** (1988) 111–20.
9. Desai, D. S., Tojo, K., Huang, Y. C. & Chien, Y. W., Transmucosal permeation of macromolecular drug: insulin. *Pharm. Res. Suppl.,* **3** (1986) 55S.
10. Murakami, T., Kishimoto, M., Kawakita, H., Higashi, Y., Yata, N., Amagase, H., Nojima, N. & Fuwa, T., Enhanced rectal and nasal absorption of human epidermal growth factor by the presence of absorption promoters in rats. *J. Pharm. Sci.,* **76** (1987) S85.
11. Moore, J. A., Pletcher, S. A. & Ross, M. J., Absorption enhancement of growth hormone from the gastrointestinal tract of rats. *Int. J. Pharmaceutics,* **34** (1986) 35–43.
12. Bocci, V., Corradeschi, F., Naldini, A. & Lencioni, E., Enteric absorption of human interferons α and β in the rat. *Int. J. Pharmaceutics,* **34** (1986) 111–14.

13. Paulesu, L., Corredeschi, F., Nicoletti, C. & Bocci, V., Oral administration of human recombinant interferon-α_2 in rats. *Int. J. Pharmaceutics*, **46** (1988) 199–202.
14. Illum, L., Microspheres as a potential controlled release nasal drug delivery system. In *Delivery Systems for Peptide Drugs*, ed., S. S. Davis, L. Illum & E. Tomlinson. Plenum Press, New York (1986) pp. 205–10.
15. Levy, F., Muff., R., Dotti-Sigrist, M. A. & Fischer, J. A., Formation of neutralizing antibodies during intranasal synthetic salmon calcitonin treatment of Paget's disease. *J. Clin. Endocrinol. Metab.* (in press).
16. Collen, D., Stassen, J.-M. & Larsen, G., Pharmacokinetics and thrombolytic properties of deletion mutants of human tissue-type plasminogen activator in rabbits. *Blood*, **71** (1988) 216–19.
17. Haber, E., Quertermous, T., Matsueda, G. R. & Runge, M. S., Innovative approaches to plasminogen activator therapy. *Science*, **243** (1989) 51–6.
18. Goldfarb, D. S., Gariépy, J., Schoolnik, G. & Kornberg, R. D., Synthetic peptides as nuclear localization signals. *Nature*, **322** (1986) 641–4.
19. Moore, H. H. & Kelly, R. B., Re-routing of a secretory protein by fusion with human growth hormone sequences. *Nature*, **321** (1986) 443–6.
20. Offord, R. E. & Rose, K., New protein and polypeptide derived conjugates especially containing a reporter group or cytotoxic agent linked through specific N-containing groups, *European Patent Applic.*, 87106113.1. Application date, 4 Nov. 1987.
21. Chaudhary, V. K., Fitzgerald, D. J., Adhya, S. & Pastan, I., Activity of a recombinant fusion protein between transforming growth factor type α and *Pseudomonas* toxin. *Proc. Nat. Acad. Sci., USA*, **84** (1987) 4538–42.
22. Chaudhray, V. K., Mizukami, T., Fuerst, T. R., FitzGerald, D. J., Moss, B., Pastan, I. & Berger, E. A., Selective killing of HIV-infected cells by recombinant human CD4-*Pseudomonas* exotoxin hybrid protein. *Nature*, **335** (1988) 369–72.
23. Feng, G.-S., Gray, P. W., Shepard, H. M. & Taylor, M. W., Antiproliferative activity of a hybrid protein between interferon-γ and tumor necrosis factor-β. *Science*, **241** (1988) 1501–3.
24. Murphy, J. R., Hybrid Protein, U.S. Patent, 4,675,382. 23 June 1987.
25. Capon, D. J., Chamow, S. M., Mordenti, J., Marsters, S. A., Gregory, T., Mitsuya, H., Byrn, R. A., Lucas, C., Wurm, F. M., Groopman, J. E., Broder, S. & Smith, D. H., Designing CD4 immunoadhesins for AIDS therapy. *Nature*, **337** (1989) 525–31.
26. Scott, C. F., Lambert, J. M., Goldmacher, V. S., Blattler, W. A., Sobel, R., Schlossman, S. F. & Benacerraf, B., The pharmacokinetics and toxicity of murine monoclonal antibodies and of gelonin conjugates of these antibodies. *Int. J. Immunopharmac.*, **9** (1987) 211–25.
27. Greenfield, L., Johnson, V. G. & Youle, R. J., Mutations in diphtheria toxin separate binding from entry and amplify immunotoxin selectivity. *Science*, **238** (1987) 536–9.
28. Taetle, R., Honeysett, J. M. & Houston, L. L., Effects of anti-epidermal growth factor (EGF) receptor antibodies and an anti-EGF receptor recombinant-recin A chain immunoconjugate of growth of human cells. *J. Nat. Cancer Inst.*, **80** (1988) 1053–9.
29. Baenziger, J. U., The role of glycosylation in protein recognition. *Am. J. Physiol.*, **121** (1985) 382–91.

30. Miyazono, K. & Heldin, C.-H., Role for carbohydrate structure in TGF-β1 latency. *Nature,* **338** (1989) 158–60.
31. Madiyalakan, R., Chowdhary, M. S., Rana, S. S. & Matta, K. L., Lysosomal-enzyme targeting: The phosphorylation of synthetic D-mannosyl saccharides by UDP-*N*-acetylglucosamine: lysosomal-enzyme *N*-acetylglucosamine-phosphotransferase from rat-liver microsomes and fibroblasts. *Carbohydrate Res.,* **152** (1986) 183–94.
32. Kelm, S. & Schauer, R., The galactose-recognizing system of rat peritoneal macrophages. Receptor-mediated binding and uptake of glycoproteins. *J. Biol. Chem.,* **267** (1986) 989–98.
33. Wilson, G., Eidelberg, M., & Michalak, V., Selective hepatic uptake of synthetic glycoproteins. Mannosaminated ribonuclease A dimer and serum albumin. *J. Gen. Physiol.,* **74** (1979) 495–509.
34. Donahue, R. E., Wang, E. A., Kaufman, R. J., Foutch, L., Leary, A. C., Witek-Giannetti, J. S., Metzger, M., Hewick, R. M., Steinbrink, D. R., Shaw, G., Kamen, R. & Clark, S. C., Effects of N-linked carbohydrate on the in vivo properties of human GM-CSF, *Cold Spring Harbor Symposia on Quantitative Biology,* Vol. LI. Cold Spring Harbor Laboratories, Cold Spring Harbor, NY, 1986, pp. 685–92.
35. Clark, S. C., Wong, G. G. & Donahue, R. E., Colony stimulating factors having reduced levels of carbohydrate, Int Patent Number: WO 88/05786. 11 August 1988.
36. Hagen, F. S. & Kaushansky, K., Colony-stimulating factor derivatives. Eur. Patent. Appl., Publ. No., EP 0 276 846 A2. Filing date 28 January 1988.
37. Vera, D. R., Krohn, K. A., Stadalnik, R. C. & Scheibe, P. O., Tc-99m-galactosyl-neoglycoalbumin: in vivo characterisation of receptor-mediated binding to hepatocytes. *Radiology,* **151** (1984) 191–9.
38. Akiyama, A., Bednarski, M., Kim, M.-J., Simon, E. S., Waldmann, H. & Whitesides, G. M., Enzymes in organic synthesis. *Chem. Brit.* (1987) 645–54.
39. Berger, E. G., Greber, U. F. & Mosbach, K., Galactosyltransferase-dependent sialylation of complex and endo-*N*-acetylglucosaminidase H-treated core *N*-glycans in vitro. *FEBS,* **203** (1986) 64–8.
40. Berger, E. G., Müller, U., Aegerter, E. & Strous, G. J., Biology of galactosyltransferase: recent developments. *Biol. Chem. Trans.,* **15** (1987) 610–13.
41. Blakey, D. C. & Thorpe, P. E., Effect of chemical deglycosylation on the in vivo fate of ricin A-chain. *Cancer Drug Delivery,* **3** (1986) 189–96.
42. Vitetta, E. S., Fulton, R. J., May, R. D. & Uhr, J. W., Redesigning nature's poisons to create anti-tumor reagents. *Science,* **238** (1987) 1098–104.
43. Napper, D. H. & Netschey, A., Studies of the steric stabilisation of colloidal particles. *J. Colloid Interface Sci.,* **37** (1971) 528–35.
44. Abuchowski, A., Van Es, T., Palczuk, N. C. & Davis, F. F., Alteration of immunological properties of bovine serum albumin by covalent attachment of polyethylene glycol. *J. Biol. Chem.,* **252** (1977) 3578–81.
45. Katre, N. V., Knauf, M. J. & Laird, W. J., Chemical modification of recombinant interleukin 2 by polyethylene glycol increases its potency in the murine Meth A sarcoma model. *Proc. Nat. Acad. Sci., USA,* **84** (1987) 1487–91.
46. Ho, D. H. W., Wang, C.-Y., Lin, J.-R., Brown, N., Newman, R. A. & Krakoff, I. H., Polyethylene glycol-L-asparaginase and L-asparaginase studies in rabbits. *Drug Metab. Dispos.,* **16** (1988) 27–9.

Chapter 17

PROTEINS AS THERAPEUTICS: POTENTIAL AND PROBLEMS

Nowell Stebbing

ICI Pharmaceuticals, Macclesfield, Cheshire, UK

INTRODUCTION

Many of the recombinant DNA systems of current interest for production of therapeutic proteins have been covered in preceding chapters. Progress in recent years has been rapid; genes for the first proteins of clinical interest were cloned and expressed in *Escherichia coli* in the late 1970s and in less than 10 years nine materials have entered the market including insulin, growth hormone, interferon-alpha, tissue plasminogen activator and erythropoietin. This progress is remarkable when one considers the current time scales for discovery and development of chemical therapeutic agents. Although there are several very promising new production systems for therapeutic proteins, work over several years has produced some robust and generally useful systems. For simple proteins, *E. coli* expression has the advantage of high yield. For large, complex proteins such as plasminogen activator and Factor VIII, which have extensive secondary and tertiary modifications, mammalian cell production is likely to be essential. These systems can often be applied now in predictable ways and the accumulated knowledge to date would seem to be a reasonable basis for assessing accomplishments and making some extrapolations.

Progress so far

Before recombinant DNA methods, therapeutic proteins were relatively impure and poorly characterised. Only a few materials in clinical use were structurally defined (insulin, albumin, growth hormone and

227

calcitonin) and, with the exception of insulin, the degree of purity did not reach that of chemical therapeutic agents. Hence the distinction between 'biologicals' and 'chemicals' and their different treatment by regulatory authorities. Chemical therapeutic agents are characterised by specification of the final product with some assessment of the extent and nature of contaminants. Typically the product is at least 95% pure and some control of contaminants can be achieved through the chosen route of synthesis. In contrast, classical biologicals such as vaccines, immuno-globulins and Factor VIII are characterised by the methods of manufacture. The active ingredients, often less than 1% of the final product, may be assessed only by bioassays and immunoassays. Because the active ingredients of many biologicals are proteins it is perhaps not surprising that therapeutic proteins produced by recombinant DNA techniques should be viewed, at least initially, as biologicals.

Recombinant DNA methods as such did not result in the production of highly purified therapeutic proteins. However, defined and controllable production methods together with development of substantial yields have combined in such a way that chemical criteria can and have been applied to recombinant DNA derived proteins and the distinction between biologicals and chemicals has changed. Specific new concerns arose from use of recombinant DNA techniques for production of human therapeutic agents. At the outset, it was not obvious that micro-organisms, forced to express a foreign protein in large amounts, would do so effectively and that host cell contaminants could be adequately eliminated from the product. In a number of countries, concerns associated with the technology have been addressed, from a regulatory stance, through particular mechanisms, such as the 'Points to Consider' documents issued by the FDA in the United States.[1]

Recombinant DNA methods now provide means for determining the structure of previously uncharacterised therapeutic proteins. Thus, a growing list of novel and potent agents is being made available for clinical evaluation and use. A recent survey indicates that in addition to nine products already marketed, there are 97 more in development, including hormones, lymphokines, growth factors, anticoagulants and vaccines.[2]

The Issues

Proteins represent the largest and most complex agents dealt with by medicinal chemists. A chemotherapeutic agent is generally defined by its unique structural formula. This is not the case for proteins for which there can be considerable variations in tertiary structure for any defined primary amino acid sequence. Thus, there is a need to determine the structure of recombinant DNA derived therapeutic proteins and to relate this to that of the natural counterparts. This is far from trivial because the structure of the natural material may be unknown. Moreover, the extent and precise nature of some secondary modifications, such as carbohydrate moieties, remain uncertain in most cases. Folding and conformation of proteins can be dramatically affected by the procedures involved in their purification. The high concentrations (1 mg/ml) associated with preparation of a therapeutic product for clinical use may be a million-fold greater than the normal circulating concentration of the protein.[3] This is a major factor to be considered in the development of therapeutic proteins. As we shall see, there are reasons to believe that most recombinant DNA derived proteins will differ, intentionally or otherwise, from their natural counterparts. Thus they are 'homologues' (rather than 'analogues') of the native proteins. This must raise some concerns about responses to non-natural materials. Because they are natural proteins, there was a widespread belief that materials, such as interferons, would not be toxic. Unfortunately this has not proved to be the case. Determination of the relationship between toxicity and efficacy remains an important goal towards which little progress has been made so far.

Development of specific antibodies to therapeutically active proteins has been observed with proteins now used in clinical practice.[4] In general, these responses have not significantly limited the clinical use of proteins. Moreover, totally foreign proteins have been administered for example in the form of vaccines. Adverse responses have been limited. Non-specific 'serum sickness' has been a feature of large doses of heterologous globulins but these effects have not been observed with recombinant DNA derived proteins, including antibodies.[3] Early in the production of rDNA-derived proteins there were problems associated with contaminating endotoxins. With high production levels and purification methods suitable for parenteral agents, these problems have been overcome.

Several mammalian cell production systems have been developed and this remains an area of intense activity. For mammalian cells in culture the production costs are high, compared with bacteria and yeast, because the medium is generally complex and contains serum. Production costs are also increased by virtue of the stringent containment conditions required for such slow-growing cells. The greatest problems to be tackled so far with mammalian cell production systems, other than those relating to production costs, relate to the fact that the cells involved are transformed and thus potentially tumourigenic (see Chapter 13, this volume).

STRUCTURE AND ACTIVITY

Potency and proof of identity remain the critical features of recombinant DNA derived proteins. In many ways these features have been easier to address than for earlier biologicals. Recombinant DNA derived proteins are better characterised and generally of much higher purity than most materials derived from natural sources. Identity tests include estimates of product molecular weight, overall amino acid composition and sequence data. Determination of the pattern of tryptic fragments separated by HPLC, gives an estimate of consistency of the amino acid sequence. As for the earlier generation of biologicals, bioassays remain of great importance.

There are relatively few methods for determining secondary and tertiary structure of proteins. Such measurements are likely to be critical in terms of the consistency and potency of protein products. Physico-chemical methods such as circular dichroism, infra-red spectroscopy and analytical ultracentrifugation have been used, but not routinely. If the transformation of proteins into 'chemicals' is to be accomplished, then there is likely to be increased demands for methods for the characterisation of the secondary and tertiary structures of proteins.

Biological activity, in as much as it is sensitive to correct protein folding, if a useful indicator of structure. For proteins consisting of a single amino acid sequence with unambiguous or no disulphide bonds, it is now reasonable to assume that recombinant DNA methods can produce therapeutic proteins essentially identical to the natural material. However, there are numerous cases of differences from the natural counterparts. For example, with *E. coli* production, the retention of an N-terminal methionine or even formyl-methionine; absence of some natural N-terminal modification, such as pyro-glutamine on IFN-γ; deliberate elimination of cysteines involved in disulphide bonds, as for IFN-β and IL-2; and absence of glycosylation, such as IFN-β and IFN-γ which naturally contain about 20% carbohydrate. Moreover, *E. coli* derived materials may not undergo processing, such as naturally occurs at the C-terminus of IFN-γ. More subtle and less studied modifications are illustrated by phosphorylation at serines 132 and 142 of IFN-γ, which does not occur in *E. coli* but appear to increase biological activity.[5]

In view of these issues, it would seem appropriate to accept that many recombinant DNA derived therapeutic proteins will not be identical to their natural counterparts. Thus, greater characterisation of the structures to be used clinically would seem prudent and further emphasises the importance of viewing the new materials as chemical rather than biological agents.

SAFETY EVALUATION

Toxicology concerns initially focussed on use of production systems containing recombinant DNA, transfer of the DNA to the final product and its possible adverse effects. It is clear from the previous section that the majority of the new therapeutic proteins may be different from their natural counterparts or at least the possibility of differences should be presumed. These features still form the basis of current toxicity concerns. Preclinical studies in animals are severely limited by production of neutralising antibodies to non-homologous proteins. However, experience with the newer materials would seem to provide an adequate basis for safety evaluation of these materials.[6]

Production Methods and Materials

Sensitive immunoassays for detecting host cell contaminants have been developed. Although such assays are useful, they are not easily quantified. Independent assays for contaminating DNA have been developed using molecular hybridisation to DNA from the host production cells. Additional estimates have been sought for the possible carry-over of DNA from the production system to the final product. It has become practice to provide data from clearance studies for DNA carry-over at each production step and hence the theoretical overall carry-over into the final product. Estimates of the size of DNA in the final product may be obtained and thus the possibility of the presence of oncogenic DNA in the final product can be reasonably assessed.[7]

Routine production of a therapeutic protein typically begins from a vial of cells derived from a uniform 'master cell bank'. This master cell bank is generally extensively characterised, far in excess of current requirements for production of biologicals. However, the assays involved must help to decrease the chances of transferring contaminants into the final product. Current requirements for mammalian cells include bacteriological and virological assays, particularly for retroviruses, and direct tumourigenicity studies in animals. Such studies have now been extensive for at least three cell lines, namely CHO, VERO and mouse C127 cells, and these three systems, together with hybridoma cells, constitute the principal cellular production systems for complex proteins currently in clinical practice or trials (see Chapter 6, this volume).

Preclinical Studies

Extensive preclinical studies in various species remains a feature of the development of chemical therapeutic agents and rightly so in view of the fact that living systems may never have been exposed to many of the agents produced. In contrast, many biological agents are essentially the materials which our bodies produce or are exposed to (in the case of vaccines). Thus, in general, biologicals and recombinant DNA derived protein therapies have been subjected to limited preclinical animal studies prior to clinical use, typically 14 day studies in two species. Specificity of pharmacological effects and toxicity have been and remain important guides towards the discovery of chemotherapeutic agents. In addition, pharmacokinetic parameters have been extensively studied. This has not generally occurred with recombinant DNA derived materials.[8] However, there would seem to be good reasons to redress this difference. Nevertheless, it is worth noting that specificity is often not a feature of biological agents. We may use insulin clinically for the purpose of glucose control. However, insulin has several important metabolic effects over different concentration ranges.[9]

Unfortunately we remain largely ignorant of the features of proteins that are associated with immunogenicity.[10] It is apparent that single amino acid substitutions in a protein can affect immunogenicity[11] and this needs to be borne in mind. However, in the cases examined so far, recombinant DNA derived materials have shown immune responses comparable although qualitatively different in some cases from those of their earlier biological counterparts.[12] Despite the limited time before stimulation of antibody production, short term pharmacological and toxicological effects can be observed, with proteins, as with chemical agents. Although many interferons showed marked species specificity, this has not occurred with other agents. Erythropoietin and colony stimulating factors have biological effect in hamsters, dogs and rats as well as in monkeys. Some human interferons have also shown biological effects in laboratory species, such as hamsters. In this case a range of effects could be studied preclinically on a short-term basis.[13] Knowing whether or not a therapeutic agent has limited or extensive pharmacological and toxicological effects provides valuable guidance in clinical development. In view of the general advantage of specificity of action with chemo-therapeutic agents, it is noteworthy that erythropoietin and colony stimulating factors have shown little toxicity, in preclinical as well as clinical studies (see Chapter 15, this volume). These issues would seem to warrant attention if we are to view and develop the newer proteins more as chemical than biological agents.

The critical nature of animal studies may be illustrated by the absence of significant biological effects of erythropoietin unless it is correctly glycosylated.[14] In rat bone marrow assays the glycosylated and non-glycosylated forms of erythropoietin are equally active and the difference in activity is only apparent in animal studies. However, the reason why glycosylation is important remains unclear. In the case of glycosylated interferons and IL-2, the non-glycosylated forms produced in *E. coli* are essentially as active as their naturally glycosylated forms. Pharmacokinetic parameters also are not readily interpreted: there are examples of interferons which show little or no circulating levels, by any route of administration, which are more active than related materials which give sustained blood levels. We are still at an early stage in developing novel protein therapies and this should indicate the desirability for more rather than fewer preclinical studies, until general issues emerge.

POTENTIAL OF PROTEIN THERAPIES

There can be little doubt that the potential of protein therapies is limited by the absence of oral activity, and therefore ways of avoiding multiple injections would greatly augment clinical use. Novel delivery methods have been achieved for certain peptides (see Chapter 16, this volume). For example, a ten amino acid LHRH agonist (Zoladex) can be delivered continuously over a 1 month period from a biodegradable lactide glycolide copolymer matrix.[15, 16] Calcitonin has also been administered successfully by the nasal route.[17] Thus, discovery of active fragments of therapeutic proteins could provide possibilities for increased potential.

Despite some early predictions that real clinical utility of recombinant DNA derived proteins would be their use in devising low molecular weight analogues,[18] this has not occurred so far. Some indications of the difficulties that may be encountered are illustrated by the prolonged and contorted path that led to the first orally active angiotensin converting enzyme inhibitor: the development from an active peptide to a low molecular weight moiety was considerable.[19] However, recombinant DNA production of proteins also holds possibilities by providing novel targets for classical medicinal chemistry approaches.

Homologues and Fragments

So far, attempts to modify biological activity of recombinant DNA derived proteins by amino acid substitutions have been few in number, or have met with limited success. Although some crude structure/activity features have been determined in this way, predictive structure/activity relationships have not been established.[3] However, some interesting approaches have been adopted that point towards new proteins with novel properties. For example, hybrids have been formed between distinct proteins, such as one between IFN-γ and tumour necrosis factor, in this case with enhanced anti-proliferative activity.[20]

Prolonged activity of insulin has been achieved by various substitutions that raise the isoelectric point. The prolonged activity seems to occur because the novel homologues precipitate when they encounter the neutral pH of the body.[21] This work indicates ways in which protein modifications may be exploited for specific useful purposes.

Antibodies are rather large and complex and as yet there are no reliable methods for producing monoclonal human antibodies. However, the specificity of mouse monoclonal antibodies can be transferred into a framework of human antibodies.[22] Antibodies also represent clear examples of large proteins in which particular properties are known to be associated with certain domains, even after isolation by enzymatic cleavage from intact antibodies. The antigen binding domain is formed from the variable region of both the heavy and light chains of the tetramers forming intact antibodies. These fragments can be produced in active form in *E. coli* by secretion into the periplasmic space.[23, 24] Thus, effective antigen binding entities can be produced which are about one third the size of intact antibodies. Expression in *E. coli* should readily allow large-scale production of defined materials and the ease of gene modifications in *E. coli* holds forth the possibility of rapid generation of new antigen binding entities with altered specificity or affinity. A further development of these approaches to production of human antibody domains is the design of single-chain antigen-binding proteins.[25] Heavy and light chain variable regions may be combined through a linker peptide, designed so as to allow correct folding of the normally separate chains.

Despite considerable effort no active peptides mimicking regions in recombinant DNA derived proteins have been found for some of these materials, such as the interferons and IL-2. A few residues may be omitted from the ends but essentially the entire sequence is required for activity. In the case of IFN-α there is a claim that a large fragment (residues 1–110) retains 30% anti-viral activity but this is open to question.[3] However, various amino acid substitutions may moderately increase or decrease activity. Elimination of cysteine residues involved in disulphide bonds may dramatically, but unsurprisingly, decrease activity.

Recently there have been examples of biological activity from fragments of interleukin-Iβ and other proteins. Interleukin-Iβ has a number of biological activities, including augmentation of the perception of pain (hyperalgesia) and this property can be achieved with a synthetic tripeptide (Lys–Pro–Thr), which occurs as residues 193–195 in the intact protein.[26] Substitution of the proline with D-proline results in a tripeptide with antagonist activity. Other effects of IL-Iβ are not achieved or antagonised by either of the tripeptides. A nonapeptide corresponding to residues 163–171 of IL-Iβ shows, qualitatively and quantitatively, several of the immunostimulatory properties of the intact protein without the inflammation and shock effects.[27] Another recent example of full biological activity residing in a peptide found as a defined internal sequence is a heptapeptide corresponding to residues 247–253 of lipocortin I.[28] In this case, activity is associated with the terminal residues of the peptide since the internal sequence lacks activity. The heptapeptide is also found in another anti-inflammatory protein (uteroglobin) and biological activity has been observed *in vivo* in the carrageenan-induced oedema model as well as by inhibition of phospholipase A_2. The potency of the anti-inflammatory effect of the peptide (10× greater than indomethacin) is remarkable. These studies indicate that multiple activities may reside in specific amino acid sequences of pluripotent proteins and provide leads to novel therapeutic agents.

Proteins as Targets

Recombinant DNA methods are now revealing and allowing production of receptors and the molecular transducing agents that mediate changes in a wide range of pharmacological effects.[29] Knowledge and availability of these materials provides valuable guides to classical chemotherapy approaches. The simple knowledge of the number of different types of GABA receptors or adrenergic receptors may indicate the possibilities and limits for specific agents. Moreover, recombinant DNA methods allow investigation of the structural features of receptors themselves and this should provide additional information in drug design.[30] Thus we may expect a considerable extension of medicinal chemistry approaches based on the availability of new protein targets that are now available through recombinant DNA methods (see also Chapter 14, this volume).

Modification of Protein Production

A more fundamental approach to affecting processes mediated by proteins would be control of production of the protein itself. This extension of protein therapies requires an understanding of the molecular processes that are inadequate or aberrant in specific diseases. A further extension of this approach leads to gene therapy: the specific introduction or deletion of genes necessary to overcome deficiency diseases or diseases associated with over-production of certain materials. These extensions are really beyond the scope of this volume.

FURTHER DEVELOPMENTS

Although there are only a few therapeutic proteins in medicine today, there is no reason to believe that the number of potentially useful materials is limited. As we have seen, application of recombinant DNA methods has added several more useful proteins to clinical medicine, beyond the four derived from earlier methods, and more have been identified, including natural ligands for the benzodiazepine receptors[31] and cardiac mediators of blood pressure control.[32] Isolation of the calcitonin gene has led to identification of a related protein with distinct activity as a potent vasodilator.[33] Moreover, consideration of the human genome would indicate that there may be several thousand gene based diseases. While not all of these will be amenable to therapy, the number is likely to be considerable for direct protein therapy or treatments based on protein targets, as outlined here. As indicated by this volume, production methods are developing very rapidly, even for complex proteins. These considerations should encourage us to believe that there is much more to come from recombinant DNA methods than we have seen so far, even for therapeutic proteins.

While there is much that forms a basis for optimism for new protein based therapies, there are also significant immediate puzzles that need to be resolved for a greater understanding of the molecular basis of many diseases. In several cases it is apparent that disease arises not from a defective gene coding for a protein but from a segment of DNA that controls production. We also do not understand why some proteins are produced as a family of related structures (IFN-αs, growth factors) while others are produced as homogeneous materials from single genes (IFN-γ, erythropoietin). In developing new protein therapies, potency is not generally the critical issue. Specificity of action, which we prize so highly in therapy, is a significant problem in some cases, as is protein delivery. These associated problems represent major challenges for the next phase of protein therapies.

REFERENCES

1. Fenno, J., Points to consider about "Points to Consider". *Medical Device and Diagnostic Industry,* **8** (1986) 8–9.
2. Mossinghoff, G. J., *Update: Biotechnology Products in Development.* Pharmaceutical Manufacturers Association, Washington, 1988.
3. Stebbing, N., Biotherapeutic agents obtained by recombinant DNA methods. In *Principles of Cancer Biotherapy,* ed. R. K. Oldham. Raven Press, New York, 1987, pp. 65–91.

4. Ross, J. M., Allergy to insulin. *Pediatr. Clin. North Amer.,* **31** (1984) 675–87.
5. Arakawa, T., Parker, C. G. & Lai, P. H., Sites of phosphorylation in recombinant Human Interferon-γ. *Biochem. Biophys. Res. Commun.,* **136** (1986) 679–84.
6. Stebbing, N., Risk assessment for recombinant DNA-derived proteins. In *Banbury Report 29: Therapeutic Peptides and Proteins: Assessing the New Technologies,* ed. D. R. Marshak & D. T. Liu. Cold Spring Harbor Laboratory Press, New York, 1988, pp. 189–99.
7. Hopps, H. & Petricciani, J. (eds) *Abnormal Cells, New Products and Risk.* Tissue Culture Association, Gaithersburg, 1985.
8. Stebbing, N., Development of biological agents through DNA technology. In *The Focus of Pharmaceutical Knowledge: Proceedings 6th International Meeting of Pharmaceutical Physicians,* ed. D. R. Burley, B. Mulliger & C. Harward. The Macmillan Press, London, 1988.
9. Parsons, J. A., Endocrine pharmacology. In *Peptide Hormones,* ed. J. A. Parsons. McGraw-Hill, New York, 1979, pp. 67–84.
10. Benjamin, D. C., Berzofsky, J. A., East, I. J., Gurd, F. R. N., Hannum, C., Leach, S. J., Margoliash, E., Michael, J. G., Miller, A., Prager, E. M., Reichlin, M., Sercarz, E. E., Smith-Gill, S. J., Todd, P. E. & Wilson, A. C., The antigenic structure of proteins: a reappraisal. *Ann. Rev. Immunol.,* **2** (1984) 67–101.
11. Ibrahimi, I. M., Eder, J., Prager, E. M., Wilson, A. C. & Arnon, R., The effect of a single amino acid substitution on the antigenic specificity of the loop region of lysozyme. *Molec. Immunol.,* **17** (1980) 37–46.
12. Chiu, Y.-Y. H. & Sobel, S., Scientific review on the safety of peptide hormones: insulin, growth hormone, and LHRH and its analogs. In *Banbury Report 29: Therapeutic Peptides and Proteins: Assessing the New Technologies,* ed. D. R. Marshak & D. T. Liu. Cold Spring Harbor Laboratory Press, New York, 1988, pp. 219–27.
13. Stebbing, N., Downing, M., Fagin, K. D., Altrock, B. W., Fish, E. N., Liggitt, H. D., Moochhala, S. & Renton, K., Experimental animal models for assessing clinical utility of human interferons: studies in hamsters. In *The Biology of the Interferon System.* ed. W. E. Stewart & H. Schellekens. Elsevier Science Publishers, Leiden, 1986, pp. 411–18.
14. Browne, J. K., Cohen, A. M., Egrie, J. C., Lai, P. H., Lin, F. K., Stickland, T., Watson, E. & Stebbing, N., Erythropoietin: gene cloning, protein structure and biological properties. *Cold Spring Harbor Symp. Quant. Biol.,* **51** (1987) 693–702.
15. Ahmed, S. R., Grant, J., Shalet, S. M., Howell, A., Chowdhury, S. D., Weatherson, T. & Blacklock, N. J., Preliminary report on use of depot formulation of LHRH analogue ICI 118630 (Zoladex) in patients with prostatic cancer. *Brit. Med. J.,* **290** (1985) 185–7.
16. Furr, B. J. A. & Jordan, V. C., The pharmacology and clinical uses of Tamoxifen. *Pharmacol. Ther.,* **25** (1984) 127–205.
17. Thamsborg, G., Storm, T., Sykulski, R., Brinch, E., Andersen, N. F., Holmegard, S. N. & Soerensen, O. H., Intransasol calcitonin and prevention of postmenopausal bone loss. *Lancet,* **1** (1988) 413.
18. Vane, J. & Cuatrecases, P., Genetic engineering and pharmaceuticals. *Nature,* **312** (1984) 303–5.

19. Horovitz, Z. P., In *Parmacological and Biochemical Properties of Drugs, Vol. 3,* ed. M. E. Goldberg. American Pharmaceutical Association, 1981, pp. 148–75.
20. Feng, G. S., Gray, P. W., Shepard, H. M. & Taylor, M. W., Antiproliferative activity of a hybrid protein between interferon-γ and tumor necrosis factor-β. *Science,* **241** (1988) 1501–3.
21. Markussen, J., Diers, J., Engesgaard, A., Hansen, M. T., Hougaard, P., Langkjaer, L., Norris, K., Ribel, U., Sorensen, A. R., Sorensen, E. & Voigt, H. O., Soluble, prolonged-acting insulin derivatives. II. Degree of protection and crystallizability of insulins substituted in positions A17, B8, B13, B27 and B30. *Protein Engng,* **1** (1987) 215–23.
22. Verhoeyen, M., Milstein, C. & Winter, G., Reshaping human antibodies: grafting an antilysozyme activity. *Science,* **239** (1988) 1534–6.
23. Skerra, A. & Pluckthun, A., Assembly of a functional immunoglobulin Fv fragment in *Escherichia coli. Science,* **240** (1988) 1038–41.
24. Better, M., Chang, C. P., Robinson, R. R. & Horwitz, A. H., *Escherichia coli* secretion of an active chimeric antibody fragment. *Science,* **240** (1988) 1041–3.
25. Bird, R. E., Hardman, K. D., Jacobson, J. W., Johnson, S., Kaufman, B. M., Lee, S.-W., Lee, T., Pope, S. H., Riordan, G. S. & Whitlow, M., Single-chain antigen-binding proteins. *Science,* **242** (1988) 423–6.
26. Ferreira, S. H., Lorenzetti, B. B., Bristow, A. F. & Poole, S., Interleukin-Iβ as a potent hyperalgesic agent antagonised by a tripeptide analogue. *Nature,* **334** (1988) 698–700.
27. Boraschi, D., Nencioni, L., Villa, L., Lensini, S., Bossu, P., Ghiara, P., Presentini, R., Perin, F., Frasca, D., Doria, G., Forni, G., Musso, T., Giovarelli, M., Ghezzi, P., Bertini, R., Besedovsky, H. O., Del Rey, A., Sipe, J. D., Antoni, G., Silvestri, S. & Tagliabue, A., In vivo stimulation and restoration of the immune response by the noninflammatory fragment 163–171 of human interleukin Iβ. *J. Exp. Med.,* **168** (1988) 675–86.
28. Miele, L., Cordella-Miele, E., Facchiano, A. & Mukherjee, A. B., Novel anti-inflammatory peptides from the region of highest similarity between uteroglobin and Lipocortin I. *Nature,* **335** (1988) 726–9.
29. Lester, H. A., Heterologous expression of excitability proteins: route to more specific drugs? *Science,* **241** (1988) 1057–63.
30. Escobedo, J. A. & Williams, L. T., A PDGF receptor domain essential for mitogenesis but not for many other responses to PDGF. *Nature,* **335** (1988) 85–9.
31. Ferrero, P., Guidotti, A., Conti-Tronconi, B. & Costa, E., A brain octadeca-neuropeptide generated by tryptic digestion of DBI (diazepam binding inhibitor) functions as a pro-conflict ligand of benzodiazepine recognition sites. *Neuropharmacology,* **23** (1984) 1359–62.
32. Lang, R. E., Tholken, H., Ganten, D., Luft, F. C., Ruskoaho, H. & Unger, T., Atrial natriuretic factor — a circulating hormone stimulated by volume loading. *Nature,* **314** (1985) 264–6.
33. Brain, S. D., Williams, T. J., Tippins, J. R., Morris, H. R. & MacIntyre, I., Calcitonin gene-related peptide is a potent vasodilator. *Nature,* **313** (1985) 54–6.

INDEX

Acetolactate synthetase, 164
Adenosine deaminase gene, 84
Administration routes for
    therapeutic proteins, 211–13
AIDS, 197, 201
Airlift fermentation, 172–5
Albumin secretion, 66–9
Alfalfa mosaic virus, 161
Alpha-factor mating pheromone,
    51–3
AMI. *See* Heart disease
Anchorage-dependent animal cell
    culture reactors, 175–6
cell culture media, 176
Animal cell-culture reactors, 171–6
Animals and protein production
    higher, 153–5
    mammals. *See* Mammalian cell-
    expression systems
    transgenic, 150–3
Antibiotics, 15, 234
Anti-EGF, 216
Antigen production, 30–1
    vaccinia, and, 106–8
Arabidopsis, 164, 167

Bacillus subtilis, 1–14
    chromosomal integration, 6–7
    controlled heterologous proteins,
    expression, 7–9

Bacillus subtilis—contd.
    stable heterologous protein
    expression, 4–7
Bacillus thuringiensis crystal toxin,
    165
Baker's yeast fermentation, 63–4
Blood clotting factors. *See*
    Recombinant factor VIII
Bone-marrow transplants, 196–7
Bovine papilloma virus vectors, 87–9
Bromoxynil, 165

Cancer and haemopoietic growth
    factors, 195–7
Cauliflower mosaic virus, 160
Chemotherapy. *See under*
    Haemopoietic growth factors
Chinese hamster ovary (CHO) cells
    and cloned genes, 81–7, 150
Chromosomes, *B. subtilis*, 6–7
Cloned genes, 79–98
    CHO cells, 84–7
    expression systems, 81–91
    ideal expression systems, 91–4
    murine myeloma cell lines, 90–1
    vectors, BPV, 87–9
Colony-stimulating factor. *See*
    under Haemopoietic growth
    factors
Crabtree effect, 64–5

239